Routledge Revivals

A Century of Weather Service

First Published in 1970, *A Century of Weather Service* provides a comprehensive history of the birth and growth of the National Weather Service from 1870 to 1970 in America. It discusses important themes such as coping with disaster; American weather pioneers; a military weather service; The United States Weather Bureau; the air commerce age; weather in war; growth of a global weather service; calculated weather risks; the air we breathe; and one world of weather. The book closes with a chronology of the meteorological milestones of the American weather services from 1644 to 1970.

This is an important historical work for students of environmental geography and general readers interested in the topic.

T0361865

A Century of Weather Service

A History of the Birth and Growth of the National Weather Service, 1870-1970

Patrick Hughes

Routledge
Taylor & Francis Group

First published in 1970
by Gordon & Breach Science Publishers, Inc.

This edition first published in 2024 by Routledge
4 Park Square, Milton Park, Abingdon, Oxon, OX14 4RN

and by Routledge
605 Third Avenue, New York, NY 10017

Routledge is an imprint of the Taylor & Francis Group, an informa business

© 1970 by Gordon & Breach Science Publishers, Inc.

Publisher's Note
The publisher has gone to great lengths to ensure the quality of this reprint but points out that some imperfections in the original copies may be apparent.

Disclaimer
The publisher has made every effort to trace copyright holders and welcomes correspondence from those they have been unable to contact.

A Library of Congress record exists under LCCN: 78107947

ISBN: 978-1-032-86283-5 (hbk)
ISBN: 978-1-003-52219-5 (ebk)
ISBN: 978-1-032-86284-2 (pbk)

Book DOI 10.4324/9781003522195

A CENTURY OF WEATHER SERVICE

A History of the Birth and Growth of the National Weather Service

1870 - 1970

PATRICK HUGHES

Environmental Data Service
Environmental Science Services Administration
U. S. Department of Commerce
Washington, D. C.

GORDON AND BREACH, SCIENCE PUBLISHERS, INC.

NEW YORK   LONDON   PARIS

FOREWORD

February 9, 1970 will mark the 100th anniversary of the founding of our national weather service. In 1870 much of the country was still wilderness, and the telegraph was the most advanced means of communication. Today, we have turned from taming the land to tackling the problems of our environment, and television brings us live pictures from the moon. The parallel growth and organizational diversification of our Nation's weather services to meet the challenges of this century of radical change have contributed importantly to our country's emergence as a world leader in scientific thought and application.

I salute the men and women of the Nation's weather services on their centennial. May their second century of service be as beneficial to the American people as has been the first!

MAURICE H. STANS
Secretary of Commerce
U.S. Department of Commerce
Washington, D.C.

INTRODUCTION

The Weather Bureau, through its forecasts and warnings, probably touches the daily lives of more Americans than all other Government agencies combined, with the single exception of the Post Office. Yet, the Weather Bureau is itself only one of many agencies which collectively comprise a national weather service whose roots reach back to colonial days.

Only a century ago weather warnings and forecasts were not available, whether for farmer or President, and the atmosphere was still largely a realm of mystery. In the application of science to free man from the limitations of his environment, the past 100 years have far exceeded the previous hundreds of thousands. And from the beginning of this century, American weather scientists have been among the first to apply the latest technological innovation to the needs of the Nation. More recently, they have turned their attention to the needs of the global community.

ACKNOWLEDGMENTS

In a work such as this the author is often more craftsman than creator, chipping and polishing pieces provided by others to produce an overall mosaic. That the picture is as complete and detailed as it is, is due in large measure to the contributions and efforts of many people. It would be impossible to credit all who have contributed, but some must be mentioned.

Material and cooperation were provided in abundance by Gail L. Bradshaw, public information officer for the Atomic Energy Commission, and David Slade, an Environmental Science Services Administration (ESSA) meteorologist working with the AEC; Robert McCormick, an ESSA meteorologist attached to the Department of Health, Education, and Welfare's National Air Pollution Control Administration; Gordon Webb, Dr. Arnaud J. Loustalot, William White, and Dr. Wayne D. Rasmussen of the Department of Agriculture; Donald J. Byers of the Federal Aviation Agency's Office of Information Services; Roy E. Oltman and Donald M. Thomas of the Department of the Interior's Geological Survey; Michael L. Garbacz, Joseph McRoberts, and Donald Zylstra of the National Aeronautics and Space Administration; Frances Whedon of the Army's Environmental Sciences Division; Chief Warrant Officer Joseph Greco of the Coast Guard; Waldo E. Smith, Executive Director of the American Geophysical Union; and Alfred G. Kildow, Director of Public Affairs for the American Institute of Aeronautics and Astronautics.

The contributions of J. J. Keyser of the Naval Weather Service Command, Charles Dickens, Air Weather Service historian, and Carl Posey of ESSA's Public Information Office were particularly valuable. Mr. Keyser and Mr. Dickens provided the bulk of the material on the military weather services in Part III, while Mr. Posey contributed substantially to portions of Part IV.

Although the early and recent history of the Nation's weather services is well documented, information for the years between is often sketchy, scattered and, in many instances, not available in print. Here the memories and knowledge of several eminent scientists filled the void. Dr. Francis W. Reichelderfer, an architect of the Navy's Meteorological Service and for many years Chief of the Weather Bureau, Dr. Helmut E. Landsberg, former Director of the Bureau's Office of Climatology, and Dr. Charles W. Abbot, Secretary Emeritus of the Smithsonian Institution, recalled the spirit as well as the accomplishments of years past. Unpublished papers and personal correspondence provided by Albert Showalter of the Weather Bureau also supplied background material for some of the notable events of this period.

Though the chronology at the back of the book is drawn from many sources, the lists of meteorological milestones prepared by Malcolm Rigby of ESSA were the principal references, as they were in the early structuring of the narrative. Mr. Rigby also provided notes on the history of the American Meteorological Society and an outline of developments in international meteorology. In addition, he served as a general reviewer and critic and, with the help of his wife Marian, prepared the index.

The tireless efforts of William West and Clyde Collier of ESSA, of Maurice Callahan, assistant archivist of the Smithsonian Institution, and of Commander Thomas Fredian and Lieutenant David Sokol of the Naval Weather Service Command, made it possible to assemble an outstanding collection of photographs. Thanks to their efforts, and the generous cooperation of the many agencies, organizations, and individuals who supplied pictures, only the final selection was left to the author.

James Osmun of ESSA, Executive Secretary of the Weather Services Centennial Steering Committee, provided the contacts, coordination, and initial material needed to begin the book. Mary Ellis Moore of ESSA's Public Information Office was a tireless girl Friday during the final hectic months, providing invaluable assistance to the author by coordinating the efforts and activities of all connected with the project.

The typing, retyping, and retyping was done with great patience and perseverance by Catherine Shaver and Lillian Hovermale of the Environmental Data Service, who also took it upon themselves to catch errors surviving the editorial process.

Carol Litman of the author's own office was the chief editorial critic and kept him within reasonable range of established English grammar and usage. Gertrude Fricke of the same office undertook much of the tedious checking of details and references so necessary to a work of this nature.

To these, and to the many others I have not mentioned, either through oversight or lack of space, my thanks; without their efforts and contributions there would have been no book.

PATRICK HUGHES

CONTENTS

PART I
AN AMERICAN ORIGINAL

An American Original

When Central Pacific's "Jupiter" and Union Pacific's Engine No. 119 touched cowcatchers at Promontory Point, Utah on May 10, 1869 large areas of the West were still uninhabited by white men. Twenty years later the great transcontinental railroads had peopled these vast open spaces, and the American Frontier had disappeared forever.

During these same years the telephone and incandescent light bulb were invented, and at their close in 1890, the world's first skyscraper was erected in Chicago. In this period of territorial expansion and accelerating technological innovation the American weather service was born and passed its adolescence.

A Child of the Telegraph

Worrying about the weather is as old as man himself. He has always tried to understand and predict it, and the beginnings of his weather knowledge can be traced to the earliest civilizations. However, modern meteorology—the science of weather—was born only a few hundred years ago with the invention of the thermometer and barometer. Suddenly, for the first time, it was possible to accurately measure temperature and pressure, the basic elements of weather.

Heinrich W. Brandes of the University of Breslau drew the first known weather map in 1816, using observations made in 1783. He made the major discovery that storms are simply moving systems of low pressure easily identified and tracked on weather maps.

During the next few decades the weather map was adopted by more and more scientists, but there was no way to collect *current* observations from widely scattered stations quickly enough to make weather forecasts—until the invention of the telegraph.

Joseph Henry established the Nation's first telegraph storm
warning network in 1849.   —THE SMITHSONIAN INSTITUTION

4   An American Original

The telegraph made meteorology a practical science. Weather observations from distant points could now be rapidly collected, plotted, and analyzed for pressure patterns; a series of such analyses or weather maps could be used to track the movement of storms and to predict their probable paths and speeds of advance. For the first time man had the tools he needed to observe, study, and forecast the weather. Now he had to learn to use them.

When the first commercial telegraph line opened on April 1, 1845, many people saw the possibility of "forecasting" storms by simply telegraphing ahead what was coming. The first person to do something about it was Joseph Henry, Secretary of the new Smithsonian Institution.

On December 8, 1847 Henry wrote to the Smithsonian's Regents:

> . . . It is proposed to organize a system of observation which shall extend as far as possible over the North American continent. . . . The Citizens of the United States are now scattered over every part of the southern and western portions of North America, and the extended lines of the telegraph will furnish a ready means of warning the more northern and eastern observers [weather in the eastern United States usually moves from the west or southwest] to be on the watch from the first appearance of an advancing storm.

By the end of 1849, 150 widely scattered volunteers were reporting weather observations to the Smithsonian regularly.

The next logical step was to portray the weather occurring over various sections of the country on a daily map. In 1850 Henry mounted such a map in the hall of the Institution where it could easily be seen by the public. Corresponding weather signals were displayed on the high tower of the Smithsonian castle.

Henry's system of weather observers, telegraphic relay, and weather display maps was a direct ancestor of the national weather service established 20 years later. The only major element missing was the public weather forecast.

On February 1, 1868 Professor Cleveland Abbe became director of the Cincinnati Astronomical Observatory. Six months

Cincinnati, Ohio as Cleveland Abbe knew it.
—THE NATIONAL ARCHIVES

later he sent a letter to the local chamber of commerce detailing a plan to issue daily weather reports and storm warnings and asking financial support. The chamber agreed to underwrite Abbe's weather service for a 3-month trial period.

The first *Weather Bulletin* was issued for the chamber on September 1, 1869; a little later, the bulletins were printed and distributed to the public. The first weather forecast was published in the Bulletin of September 22.

Less than 6 months after Abbe's first bulletin, a national weather service was born. The man who started the immediate chain of events leading to its creation was Professor Increase A. Lapham of Milwaukee, a student of meteorology and a weather observer for both Henry and Abbe.

To Cope with Disaster
In 1868 storms sank or damaged 1,164 vessels on the Great Lakes, killing 321 sailors and passengers. In 1869 there were 1,914 casualties with 209 lives lost.

Once again, as he often had in the past, Professor Lapham sought support for a storm warning service for the Lakes, sending clippings of the maritime casualties to General Halbert E. Paine, Congressman for Milwaukee. In an accompanying letter he asked if it were not ". . . the duty of the Government to see whether anything can be done to prevent, at least, some portion of this sad loss in the future . . .?"

Congressman Paine was aware of the importance and practicability of the service Lapham was advocating. On February 2, 1870 he introduced a Joint Congressional Resolution requiring the Secretary of War "to provide for taking meteorological observations at the military stations in the interior of the continent and at other points in the States and Territories . . . and for giving notice on the northern (Great) lakes and on the seacoast by magnetic telegraph and marine signals, of the approach and force of storms." The Resolution was passed by Congress and signed into law on February 9, 1870 by President Ulysses S. Grant.

Paine named the Secretary of War to execute the law because "military discipline would probably secure the greatest promptness, regularity, and accuracy in the required observations." The Secretary assigned the new service to the Chief Signal Of-

Professor Increase A. Lapham. His petition to Congress led to
the establishment of a national weather service.   —ESSA

ficer of the Army, Brevet Brig. Gen. Albert J. Myer, who named
it "the Division of Telegrams and Reports for the Benefit of
Commerce."

Albert Myer had entered the Army in 1854 as an assistant
surgeon. While serving in New Mexico in 1856 he noticed
Comanche Indians signaling with their lances and devised the
now familiar code of wigwag signals with flags and torches. In
1860 Myer was appointed Signal Officer of the Army, and
charged with the development of his signal system. Now he
also had to build an organization capable of providing the
Nation with an effective weather service. Fortunately, the
groundwork had already been laid.

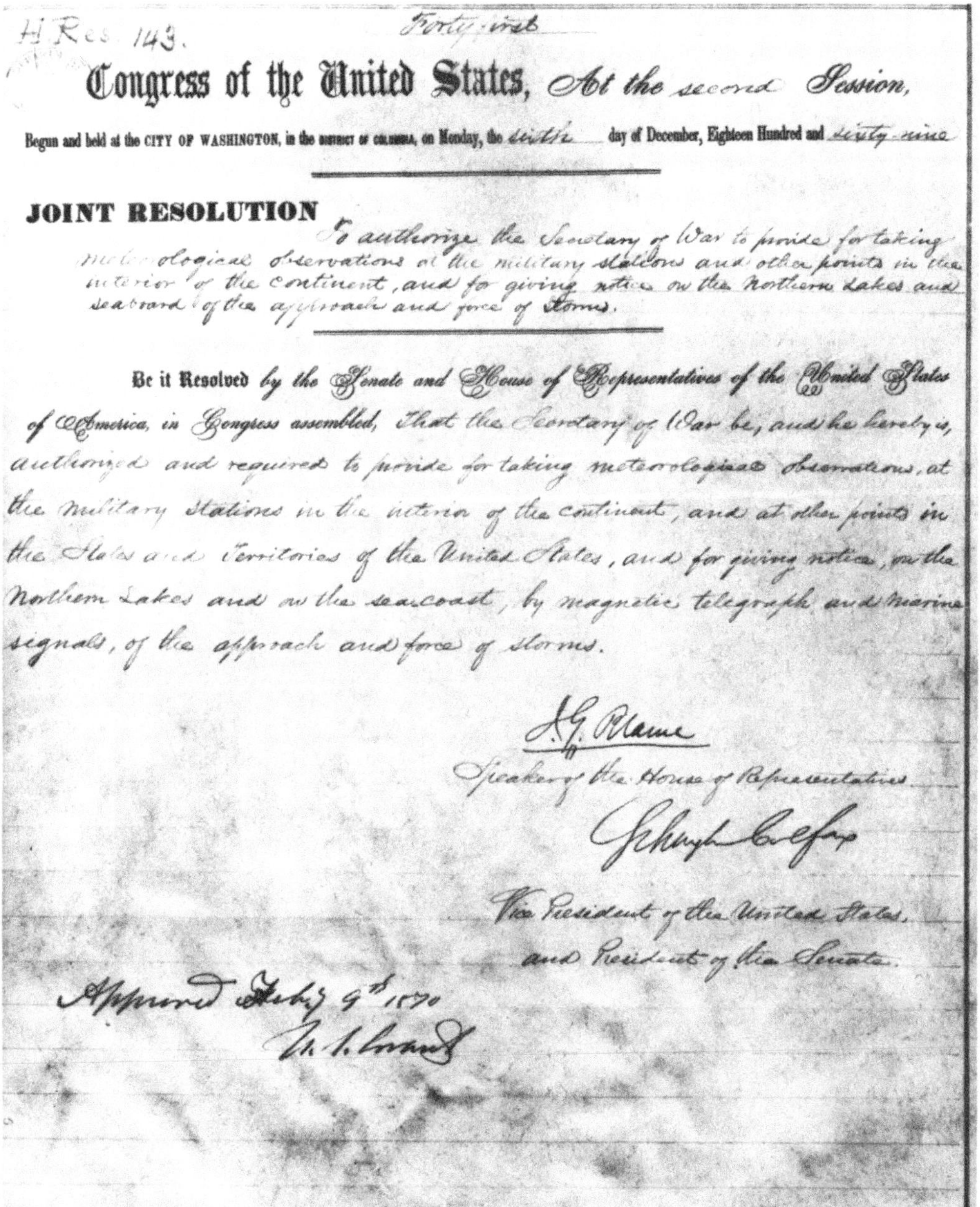

Forty-first

Congress of the United States, *At the second Session,*

Begun and held at the CITY OF WASHINGTON, in the district of columbia, on Monday, the *sixth* day of December, Eighteen Hundred and *sixty-nine*

JOINT RESOLUTION

To authorize the Secretary of War to provide for taking meteorological observations at the military stations and other points in the interior of the continent, and for giving notice on the northern Lakes and seaboard of the approach and force of storms.

Be it Resolved by the Senate and House of Representatives of the United States of America, in Congress assembled, That the Secretary of War be, and he hereby is, authorized and required to provide for taking meteorological observations, at the military stations in the interior of the continent, and at other points in the States and Territories of the United States, and for giving notice, on the Northern Lakes and on the seacoast, by magnetic telegraph and marine signals, of the approach and force of storms.

J. G. Blaine
Speaker of the House of Representatives.

Schuyler Colfax

Vice President of the United States,
and President of the Senate.

Approved Feby 9th 1870
U. S. Grant

The Joint Congressional Resolution creating the Government's first official weather service. Signed by President Grant on February 9, 1870. —THE NATIONAL ARCHIVES

An American Original 9

Brevet Brigadier General Albert J. Myer organized the new
weather service, which he established as a division of the Army
Signal Service.   —ESSA

10   An American Original

December 1799

8th. Morning perfectly clear, calm and pleasant; but about 9 oclock the wind came from the No. Wt. and blew fresh. Mer. 38 in the morning — and 30 at Night

9. Morning clear & pleasant, with a light Wind from No. W. Mer. at 33. — Pleasant all day — afternoon Calm. Mer. 39 at Night — Mr. Howell Lewis & wife set off on their return home after breakfast — and Mr. Law Lewis and Washington Custis on a Jones's & F. Tucker

10. Morning clear & calm — Mer. at 31. afternoon lowering — Mer. at 22 and Wind brisk from the Southward — A very large hoar frost this morn.

11. But little wind and Raining — Mer. 44 in the morning and 38 at Night. — About 9 oclock the Wind shifted to No. Wt. & it ceased raining, but contd. Cloudy. — Lord Fairfax, his son Thos. and daughter — Mrs. Warner Washington & son Whiting — and Mr. Jn. Herbert dined here & returned after dinner. —

12. Morning Cloudy — Wind at No. Et. & Mer. 33. — a large circle round the Moon last Night. — About 1 oclock it began to snow — soon after to Hail and then turned to a settled cold Rain — Mer. 28 at Night.

13. Morning Snowing & abt. 3 Inches deep — Wind at No. Et. & Mer. at 30. Contg. Snowing till 1 oclock — and abt. 4 it became perfectly clear — wind in the same place but not hard — Mer.

This page of George Washington's weather diary probably contains the last words he ever wrote. The final entry was made on December 13; he died on the 14th.

—THE LIBRARY OF CONGRESS

Philadelphia.

Day	h. min	temp	Day	h. min	temp		Day	h. min	temp
1.	7-0. a.m	81½	10.	8-0. a.m	75		19	8-30. a.m	79
	7-0. P.m	82		9-15.	76½			9-0.	79
2.	6-0. a.m	78		2-0. P.m	80			4-30 P.m	79½
	9-40. a.m	78		4-45.	82			8-45.	77¼
	9-0. P.m	74		6-30.	81½		20	8-30. a.m	72
3.	5-30. a.m	71½		9-30	78			8-20.	72
	1-30. P.m	76	11.	8-30 a.m	74			2-40. P.m	78¼
	8-10.	74		8-0.	76½			6-0.	78½
4.	6-0. a.m	68		9-40. P.m	75			9-0	76
	9-0.	72½	12.	7-0. a.m	72		21	8-15. a.m	75
	1-0. P.m	76		9-0.	72			11-30.	79
	9-0.	73½		8-30. P.m	72			8-0. P.m	79
5.	6-0. a.m	71¼	13.	5-30. a.m	71¼			9-15.	77
	9-0.	72		11-0.	74		22	6-0. a.m	74
	9-0. P.m	74		2-0. P.m	76			9-0.	87½
6.	5-0. a.m	74		6-45.	76			3-20. P.m	84
	9-0.	75		7-25	76			9-15.	80
	4-0. P.m	77		9-0.	72	rain	23	6-0. a.m	75
	10-0.	74	14.	6-50. a.m	73	rain		9-15.	77
7.	6-0. a.m	71		9-30.	72	rain		10-0. P.m	78
	10-0	73		5-0. P.m	71½	rain	24	5-50. a.m	73
	1-0 P.m	74		1-35.	70			9-0.	75
	3-20.	75		5-0.	69			9-40. P.m	78½
	9-30.	74		8-45.	68½		25	6-0. a.m	75
8.	5-35 a.m	75	15.	6-30. a.m	66½			9-0.	76½
	9-0.	77½		9-0.	68¼			9-0. P.m	78
	2-0. P.m	80		7-30. P.m	69¼		26	6-0. a.m	72½
	5-0.	81		9-0.	67			9-0. P.m	77½
	8-15.	80	16.	5-45. a.m	65½			9-15. P.m	79½
	9-30.	79		9-45.	68½		27	6-0. a.m	71½
9.	5-30. a.m	75		7-15. P.m	72½			11-0.	80
	9-0	77½		9-0.	71¾			11-30.	81
	6-30. P.m	81½	17.	6-0. a.m	69¾			6-0. P.m	83½
	45	79		10-0.	75	rain		7-0.	83
				9-30. P.m	74			9-30.	91½
			18.	5-30. a.m	73		28	6-30. a.m	75
				10-15.	76			9-0.	79½
				8-0. P.m	80½	1		11-0.	82
								12-0.	80
								1-0. P.m	

A page from Thomas Jefferson's *Garden Book* (a weather diary). The entries were made in Philadelphia in 1776 and include weather observations for July 4th.

Thomas Jefferson studied the Nation's climate, collecting weather records from as far west as the Mississippi River.
—THE LIBRARY OF CONGRESS

American Weather Pioneers

The necessity of studying and understanding the weather was recognized by the Founding Fathers. Benjamin Franklin and Thomas Jefferson loom almost as large in the history of American weather science as they do in the history of our country, and the last entry in George Washington's weather diary was made the day before he died.

In September 1743 a "nor'easter" hit Philadelphia, obscuring an expected eclipse of the moon. When Benjamin Franklin, America's first weather scientist, learned that the eclipse was seen by his brother in Boston (because the storm reached there later than at Philadelphia) he rightfully concluded that the storm had moved from the southwest despite the northeast winds. This novel and important concept was years ahead of Franklin's time; it wasn't until a century later that it was recognized that storms are approximately circular wind systems that move from place to place.

Thomas Jefferson bought his first thermometer while writing the Declaration of Independence and his first barometer a few days after the document was signed. Jefferson made regular weather observations at Monticello from 1772-78, and for much of the last two of these years he and the president of William and Mary College in Williamsburg took the first known simultaneous weather observations in America.

Like George Washington, Jefferson took weather observations well into his final illness. The last entry in his "Weather Memorandum Book" was made on June 29, 1826, six days before his death.

Credit for the first American weather observation network must go to the Army's Medical Department. On May 2, 1814 Dr. James Tilton, Physician and Surgeon General of the United States Army, ordered his surgeons to keep weather diaries to investigate the influence of weather and climate upon diseases. The War of 1812 then in progress prevented immediate compliance (Tilton himself, a veteran of the Revolutionary War, was in the field with the Army), and the first known diary was kept at Cambridge, Massachusetts in 1816.

By 1838 daily weather observations were being taken by Army surgeons at 13 forts, mainly in the Midwest, and by the end of the Civil War observations had been made at 143 sta-

James Pollard Espy was appointed the Government's first official meteorologist in 1842. —The Smithsonian Institution

tions. Wars and Indian raids often interfered, and reports not infrequently contained such notations as, "Owing to the threatened outbreak of the Wallapais, the rain gage was abandoned for several days. . . ."

Army Medical Department weather observations are the earliest ones we have for the American West. Although long periods of record exist for a few fixed locations, most outposts were occupied for only a few years then abandoned as the frontier pushed westward.

The most important weather observation and reporting system in the United States prior to the founding of a national meteorological service was that established by Joseph Henry and the Smithsonian Institution. To start off on the right foot, Henry consulted the weather authorities of his day, including Elias Loomis and James Pollard Espy.

Loomis was a pioneer prober of the structure and motion of storms and is credited with the original conception of isobars, the lines of equal pressure that appear on present day weather maps. He prepared a masterly summary of weather knowledge for Henry, as well as a detailed, if idealistic, plan for the Smithsonian's proposed weather system.

In 1840 James Espy came to the Surgeon General's Office to analyze and study the Army's collection of weather observations, and a year later published his "Philosophy of Storms," which earned him the title of "Storm King." Soon after the Smithsonian was established (1846) Espy, now employed by the Navy, was directed to work with Joseph Henry and the new weather service he was organizing.

On November 1, 1848, Henry and Espy distributed a circular through Congressmen and the press to recruit volunteer observers. Their appeal brought an immediate response. In 1849 Henry persuaded the telegraph companies to allot free time for the transmission of weather reports to the Smithsonian, and the following year he used these reports to construct a daily, pictorial weather map for the public, and "beneath the weather map" was soon one of the Capital's most popular meeting places.

By 1857 Henry's telegraphic weather reports were being furnished daily to the *Washington Evening Star*, and were also widely published in the newspapers of the observers' hometowns.

Lieutenant Matthew Fontaine Maury, pioneer marine meteorologist.   —ESSA

As Henry's network of volunteer observers grew, other existing systems were gradually merged with it, including several State weather services (the most prominent were those of New York, Pennsylvania, and Massachusetts). The Smithsonian network had over 600 observers just before the Civil War.

Cleveland Abbe's work in Cincinnati showed the practicability of a public weather service.   —ESSA

Henry's weather service was seriously crippled by the War. The number of southern observers was sharply reduced and did not expand when war ended. Many northern observers continued to take observations, however, and by 1869 the number of volunteers again rose to approximately 600. While Joseph Henry and his colleagues were busy on land, Lieutenant Matthew Fontaine Maury, Superintendent of the Navy's Depot of Charts and Instruments in Washington, D. C., was collecting a great wealth of weather information from the logbooks of United States men-of-war and merchant vessels. Maury used this information to prepare charts showing the winds and weather of the world's oceans, as well as favorable sailing routes, the extent of ice fields, the feeding grounds of whales, the general physical characteristics of the ocean, and many other items of interest to mariners. His work was the first comprehensive study of ocean weather, and Maury might well be called the father of marine meteorology.

Maury's charts shortened sailing times and saved international commerce some $40,000,000–$60,000,000 annually. In 1853, largely at Maury's urging, an international Maritime Congress was held in Brussels and agreement reached on a uniform system of observing and logging weather at sea. The nations present agreed to have their vessels make observations on a regular basis and routinely exchange them. This was the birth of marine meteorology as an organized science and a major milestone in international cooperation.

The Civil War abruptly ended Maury's work for the Navy; a Virginian, he fought for the Confederacy. In 1868, following an act of general amnesty, Maury was appointed professor of meteorology at the Virginia Military Institute, and spent most of the remaining years of his life promoting the use of meteorology in agriculture. He died on February 1, 1873.

Cleveland Abbe's first published weather forecast appeared in his 13th *Weather Bulletin*. It read:

> . . . Wednesday, September 22, morning cloudy and warm; afternoon, not so oppressive; evening clear and cool. Thursday, September 23, morning cloudy and warm; afternoon clear; evening clear and cool; Friday, September 24, morning clear and warm; afternoon hot.

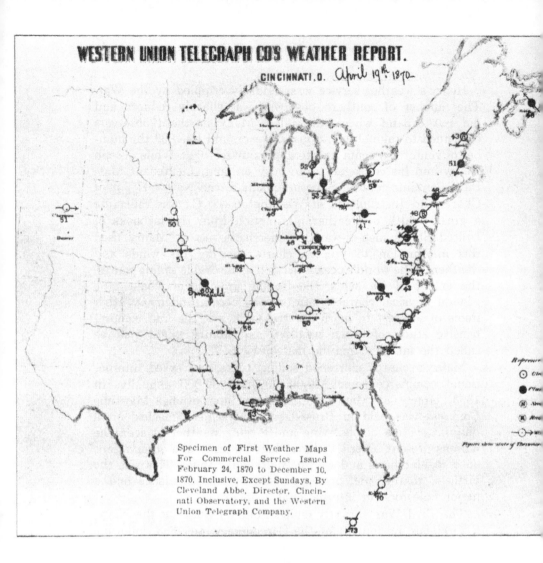

One of the daily weather reports issued by Cleveland Abbe and
Western Union for 10 months in 1870. This is not a complete
weather map, since the reports are only plotted, not analyzed.
—CINCINNATI HISTORICAL SOCIETY

Abbe's *Weather Bulletin* clearly demonstrated the practicability
of a storm warning service, and encouraged Professor Lapham
to make yet another effort to establish such a service for the
Great Lakes. And no doubt Congressman Paine, the legislative
father of the national weather service, was also familiar with
Abbe's work in Cincinnati.

A Military Weather Service

The meteorological division of the Signal Service, popularly known as the "Weather Bureau," began its work with only one asset—a widespread military telegraph system. At 7:35 a.m. on November 1, 1870 simultaneous observations were taken by 24 observer-sergeants and telegraphed to Washington and other cities and ports. With this transmission a national weather service actually came into being.

On November 8, 1870, Increase A. Lapham was appointed Assistant to the Chief Signal Officer and put in charge of the warning service for the Great Lakes. He issued the new weather service's first cautionary storm signal on the same day. Shortly after his appointment, Lapham left his position for personal reasons, and General Myer hired Cleveland Abbe to replace him. Abbe was to work for the signal service and its successor, the Weather Bureau, for over 45 fruitful years.

Forecasting the general weather was a natural outgrowth of the Signal Service's storm warning mission, and Abbe began issuing thrice-daily weather synopses and probabilities in addition to cautionary storm signals on February 19, 1871.

The Signal Service's field stations grew in number from 24 in 1870 to 284 by 1878. Three times a day each station, usually located in a principal city or town, telegraphed its observation to Washington, where the predictions were prepared and immediately telegraphed back to the observers, to railroad stations, and to the Associated Press. Eventually, the Signal Service adopted a simple flag display system to indicate the predicted local weather for the following day.

The period between 1870 and 1880 was one of rapidly expanding public service. Technically, the new agency was limited to the Great Lakes and seacoasts, but the appropriation act of 1872 authorized the provision of weather reports and signals throughout the country, "for the benefit of commerce and agriculture."

Beginning in 1873, the midnight synopses and probabilities were distributed by the field stations to thousands of rural post offices for display as "Farmers' Bulletins." This practice continued until 1881 when local signal flags replaced the bulletins. In 1880 frost warnings were issued for sugar growers in Louisiana, and the following year a system of observations and reports

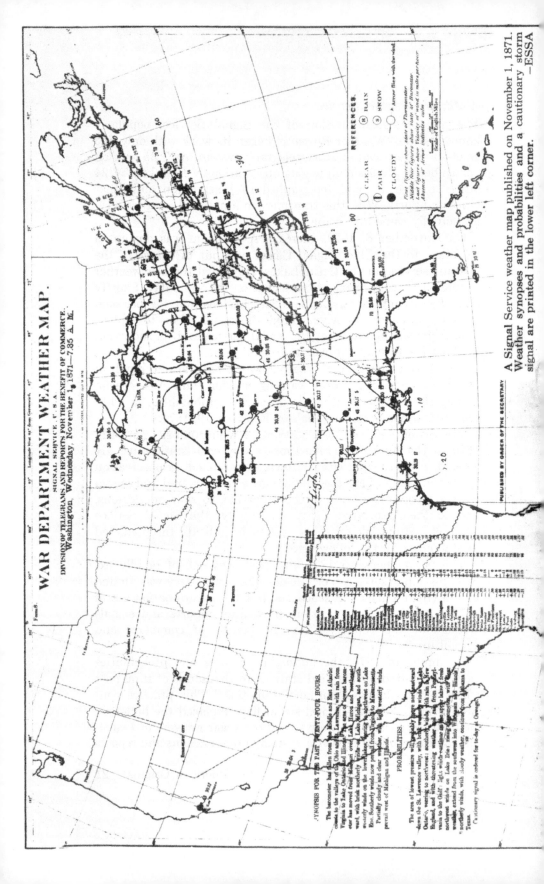

A Signal Service weather map published on November 1, 1871. Weather synopses and probabilities and a cautionary storm signal are printed in the lower left corner. —ESSA

was started for the cotton region. Later, similar specialized services were established for rice, tobacco, corn, and wheat growers.

Marine meteorological work began in June 1871 when General Myer requested that merchant ships log three daily, simultaneous weather observations, particularly when in U.S. coastal waters. The response grew gradually, and 5 years later, naval vessels were also taking such observations. By 1885 the Signal Office, in cooperation with the British meteorological service, had begun issuing warnings for Atlantic storms. In 1887 the Signal Office's marine meteorological work was transferred to the Navy's Hydrographic Office, which was continuing the work begun by Matthew Fontaine Maury.

A river stage and flood warning service was undertaken by the Signal Service in 1873. Its value was demonstrated during the Ohio Valley floods of '83 and '84, and led to the establishment of 43 special stations to monitor the Nation's rivers.

The first hurricane ever seen on a weather map appeared on the Signal Service map of September 28, 1874, and was located over the coastal waters between Savannah, Georgia and Jacksonville, Florida. The Service had earlier established six hurricane stations in the West Indies, but these had to be closed in 1881 when the legality of weather stations outside the United States was questioned. Although there were sporadic attempts to issue hurricane warnings, no systematic service was established.

In 1873 General Myer invited all nations to cooperate with the Signal Service in making and collecting one international simultaneous weather observation daily at 7:35 a.m., Washington time, at as many stations as possible throughout the world. By 1875 the Signal Office had begun daily publication of the *Bulletin of International Simultaneous Meteorological Observations*. Three years later a companion international weather map was added. These daily bulletins and maps, based on observations taken months before and mailed to Washington were used to study past weather. Their value was not fully realized until the Second World War, 65 years later. Unfortunately, the Signal Office was forced to discontinue this work in 1884 as an economy measure, although monthly publication continued until 1887.

In 1881, as part of a planned international attack on the problems of Arctic meteorology (The First Polar Year), the Signal Office dispatched two expeditions to establish northern weather outposts at Point Barrow, Alaska and Lady Franklin Bay, in the Northwest Territories of Canada. The Lady Franklin Bay expedition, led by Lieutenant Adolphus W. Greely, ended in disaster.

Greely established a weather station at Fort Conger and took meteorological and other scientific observations for two years. Then, on August 9, 1883 according to plan, he abandoned Fort Conger and brought his party to Cape Sabine, arriving October 15, 1883. Two expeditions sent to pick them up failed, however, and the party had to spend an exposed Arctic winter with few supplies. When they finally were rescued in June 1884, all but seven of the original 25 had died, and another died shortly after.

Although its emphasis was always on the practical application of weather science, the Signal Service did foster a modest meteorological research program. In the course of their studies, Cleveland Abbe and his civilian and military coworkers formulated meteorological theory, developed improved methods of forecasting and verification, studied tornadoes, cold waves, and storms in general, worked with the Chief Signal Officer in establishing international meteorological standards, developed outstanding meteorological instruments, and aided in the development of marine meteorology. In an attempt to discover the causes of storms and their laws, the Signal Service employed volunteers and professional "aeronauts" or balloonists to make upper air observations and established special weather stations at the summits and bases of mountains.

When the Signal Service was established, the Surgeon General's Office, the Smithsonian Institution, and the Corps of Engineers' Survey of the North and Northwest Lakes all had their own networks of volunteers. These observers were gradually incorporated into the weather system of the Signal Service. In 1873 Joseph Henry offered his network and records intact and during the next year Surgeon General Barnes offered to turn over the meteorological reports of the post surgeons accumulated since 1860. Myer's acceptance of both offers brought the Signal Service 383 observers from the Smithsonian system and 123 from the Army Medical Department. Unfortunately, under General

Adolphus W. Greeley led the tragic expedition to Lady Franklin
Bay in Canada's Northwest Territories.

Surgeon General Joseph K. Barnes (seated on the left) at
Camp Apache, Arizona in 1875. —The National Archives

26   An American Original

Myer the Signal Service gave little encouragement to the volunteer observers and their number decreased rapidly after 1876. General Myer died on August 24, 1880 and in December William Babcock Hazen assumed command of the recently renamed Signal Corps. Hazen, a professional soldier who had fought at Shiloh, Chickamauga, and on Sherman's march to the sea, relied largely on Greely and Abbe in meteorological matters.

When General Hazen took charge, Signal Service policy with regard to cooperative observers changed. Hazen recognized the need for voluntary reports to define the climate of the great regions of the country being opened for farming. To provide weather warnings and to collect more adequate climatological data for these regions, Hazen in 1881 approved a plan by Cleveland Abbe and Lieutenant H. H. C. Dunwoody for the organization of supplementary State Weather Services.

Such services were soon organized in Illinois, Indiana, Kansas, Michigan, Missouri, Nebraska, New Jersey, and Ohio (Iowa and Nevada already had them, but the earlier networks of New York, Pennsylvania, Ohio, and Massachusetts had been disbanded). By 1892 the country was blanketed by more than 2,000 volunteer observers belonging to State services and concerned primarily with the impact of weather on crop production.

Hazen remained Chief of the Signal Corps until he died in 1887. Following his death, Greely was appointed Chief Signal Officer by President Grover Cleveland.

Greely had entered the Signal Service some 20 years earlier and knew every man by name. During his administration, he attempted to revitalize the military (signaling) function of the Corps, but weather service remained its major concern. Greely was in charge of the Signal Corps at the termination of its civilian weather responsibilities in 1891.

The United States Weather Bureau

As early as 1880 it was clear that the War Department was not as enthusiastic over its weather service as General Myer. The Signal Corps had been almost completely absorbed by its new mission, and should its military services ever be needed, its personnel could not be spared from their weather duties.

In 1884 a joint congressional commission was set up to look into the organization of the government's various scientific bureaus. After a 2-year investigation the commission was unanimous in its opinion that the weather service did not belong in the military. Three members recommended its immediate transfer to a civilian bureau.

From 1884 to 1890 the matter was frequently debated in Congress. A bill introduced in 1887 to make Agriculture an executive department included an amendment transferring the weather service to that Department, and the Secretary of War himself requested the transfer in his *Annual Report*. In 1888 legislation to effect the transfer passed the House, but not the Senate.

By December 3, 1889, when President Benjamin Harrison again recommended transfer of the service to the Department of Agriculture, Congress was ready to act. Almost at once bills were introduced in both Houses, and on October 1, 1890, an act transferring the weather service to the Department of Agriculture was signed into law.

The work of the Signal Corps' meteorological division ended just 20 years after it began, yet, in that brief period it succeeded in establishing most of the weather services we know today. The United States already led the world in providing weather information, forecasts, and warnings to the public. As Charles Patrick Daly, a famous jurist of the day, wrote:

> Nothing in the nature of scientific investigation by the national government has proved so acceptable to the people, or has been productive in so short a time of such important results, as the establishment of the Signal Service (weather) bureau.

This was the achievement the new United States Weather Bureau inherited at noon on July 1, 1891 when the Signal Corps' weather stations, telegraph lines, apparatus, and personnel (honorably discharged) were transferred to the Department of Agriculture.

PART II
GROWING PAINS

Growing Pains

On December 8, 1903, a $50,000 "flying machine" built by Samuel Langley, Secretary of the Smithsonian, nosedived into the Potomac River in Washington, D.C., convincing most Americans that man was not meant to fly. Nine days later the Wright Brothers made their historic flights at Kitty Hawk.

The airplane and the continuing communications revolution were major factors in the fantastic growth and development of the United States through the middle decades of the 20th century. In their wake the new civilian weather service emerged as a major Government agency.

Wireless and Wings

Guglielmo Marconi began experimenting with wireless telegraphy in 1895 and proved its practicability in 1899 by transmitting across the English Channel. The Weather Bureau was the first U.S. Government agency to adopt this new and promising communications medium, and during 1901 and 1902 conducted wireless tests between two special stations at Hatteras and Roanoke Island, North Carolina.

In 1904, at the recommendation of an interagency board, President Theodore Roosevelt assigned future wireless experiments and operations on coasts and the Great Lakes to the Navy, and in the interior to the Army. The Navy was given responsibility for wireless services to meet weather needs.

Wireless telegraphy meant that the Bureau could get weather information from places where there were no telegraph wires, telephones, or cables, and from foreign countries and ships at sea. In turn, it could transmit back its forecasts and warnings. Regular radiotelegraphy broadcasts of daily weather bulletins began on July 15, 1913 from Navy radio stations at Arlington,

The wreck of Samuel Langley's *Aerodrome* in the Potomac River in 1903. — THE SMITHSONIAN INSTITUTION

The rebuilt, reinforced *Aerodrome* in flight in 1914.
—THE NATIONAL ARCHIVES

Virginia and Key West, Florida. This service was extended to the Great Lakes in 1914.

Radiotelephony appeared on the scene in 1920, an outgrowth of the development of the vacuum tube. The first systematic use made of this new medium was the broadcast of weather forecasts by the University of Wisconsin, beginning January 3, 1921. By January 1923, weather reports, daily forecasts, river forecasts, aviation forecasts, crop information, cold wave, frost, and other warnings were being broadcast by 140 radiotelephone stations in 39 states.

Beginning January 17, 1919 weather observations from 32 stations in the United States and Canada were radioed each night to the French meteorological service for immediate broadcast from the Eiffel Tower to all countries within range. European weather broadcasts to the United States began August 15, 1925.

American aviation and the Weather Bureau grew to maturity together. Bureau meteorologists provided wind data and weather statistics to pioneer heavier-than-air enthusiasts and later, after World War I had transformed flying machines from toys to transports, began issuing "flying weather" forecasts for military pilots and the newly established aerial-mail routes.

In 1926 the Air Commerce Act, "the legislative cornerstone for the development of commercial aviation in America," made the Weather Bureau officially responsible for weather services to civilian aviation ". . . to promote the safety and efficiency of air navigation in the United States and above the high seas, particularly upon the civil airways. . . ." The phenomenal growth of commercial aviation during the next few decades was the dominant factor in a similarly explosive expansion of the Weather Bureau and its services, and the catalyst for major advances in the science of meteorology itself.

Friend of the Farmer
The airplane did not really come into its own until after World War I. At first, as Congress had intended, the emphasis of the Department of Agriculture's new agency was on greater service to the farmer.

Professor Mark Walrod Harrington of the University of Michi-

Glenn L. Martin, pioneer airman, and his mother in an early
flying machine. Martin was the first aviator to take his mother
for a flight, soaring to a height of 1,000 feet over Balboa Beach,
California on May 20, 1912.—THE GLENN L. MARTIN COMPANY

gan was the first Chief of the Weather Bureau. Born and raised on an Illinois farm, Harrington was an outstanding meteorologist and climatologist and a frequent writer on the subject of weather services for agriculture.

During its first year, the civilian weather service transferred from the Signal Corps was completely reorganized and its services to the farmer greatly expanded. The number of official observation stations (which dwindled to 178 by 1890 because of economic constraint) was increased to 200, while by year's end some 3,000 volunteer observers were distributed throughout the supplementary State weather services. The additional observation posts were needed both to disseminate forecasts and storm warnings to the farmers in the field and to record the climate.

Before he became Chief, Harrington capsuled the relationship between climate and agriculture in these words:

> . . . To the farmer the problems of the climate have a permanent, while those of the weather predictions have only a passing interest. . . . It is the average weather or climate which determines the agricultural capacity of a region. An unfavorable climate is quite as vigorously exclusive as an unproductive soil. . . .

As with weather warnings and forecasts, climatology—the scientific study of climate—begins with weather observations; the difference is that climatological observations are accumulated indefinitely and the longer the period of record the clearer the picture. Professor Cleveland Abbe was an early leader in the Bureau's program of summarizing and publishing climatic information for farmers.

During the planting, growing, and harvesting season the Weather Bureau in Washington published a weekly national weather and crop bulletin. This was read by both farmer and speculator and, like the morning forecast, could determine the daily prices of farm products. This bulletin was a direct descendant of the *Weekly Weather Chronicle* first published by the Signal Service in 1872 and discontinued in 1881. Revived in 1884 as the *Weather Crop Bulletin*, it quickly inspired similar local publications by the State weather services. It has been published continuously ever since, under various names. In

1924 it acquired its current title, the *Weekly Weather and Crop Bulletin.*

By 1901 the climate and crop service of the Weather Bureau was being served by well-organized climatological services in 42 states, as well as in Cuba and Puerto Rico. These various groups were eventually organized into the Climatology Service of the Weather Bureau, for Congress, in creating the civilian weather service, had charged it with "establishing and recording the climate of the United States." By 1926 there were more than 5,000 stations publishing monthly and annual climatological summaries for the continental United States, Hawaii, the West Indies, and the Caribbean region.

The Weather Bureau also expanded the agricultural warning services initiated by the Signal Service, and added programs for cranberry, tobacco, alfalfa, truck farming, and sheep and cattle regions. Among the most useful of the services were cold wave and frost warnings.

Frost is the chief enemy of the fruit grower (in 1898 it was estimated that losses caused by frost in California annually exceeded the profits made). This special need led to the establishment of the Fruit Frost Service. The Weather Bureau was providing extensive frost warning services to Pacific Coast fruit growers by the early 1920's, and to most of the Nation by the late 1930's. In 1926 it was reported that $14 million worth of California citrus fruit was saved by warnings the Bureau issued during a single cold wave.

Expanding with the Nation
America grew rapidly during the closing years of the 19th century and the early decades of the 20th, and nowhere was this more true than on the Great Lakes.

By 1898 more tonnage was passing the port of Detroit annually than the combined total of our Atlantic, Gulf, and Pacific ports, and passenger steamers plying these waters were finer than any ocean vessel afloat. During the busy season, there was an average of 100 vessels every 24 hours, with an hourly record number of 57—nearly one a minute.

Beginning in 1895, Bureau storm warning services for Great Lakes shipping, the original impetus for a Government weather

Weather Bureau station force (including stenographer) at
Lansing, Michigan in 1898.   —ESSA

service, were intensified. By shortly after the turn of the century weather caused less than 25 percent of the marine casualties on the Lakes, and most of these involved collisions in fog,
not storms; before the Signal Service's organization in 1870,
weather was the cause of 75 percent of all marine disasters on
these waters.

By 1897, the Bureau's other basic services were also fairly
well established. In 1883 under the Signal Corps, forecasts and
warnings were distributed to some 8,000 communities and there
were 41 storm warning display stations; by 1897 over 51,000
localities were being served and there were 253 storm warning
stations—as well as some 8,000 crop and climatological reporting stations where none had existed earlier.

Americans were still moving westward, and the Rocky Mountains and Pacific Coast were now the Nation's frontiers. By
1900 the populations of New York City, Chicago, and Phila-

delphia had topped one million. The country was splitting its seams, and according to Willis Luther Moore, the colorful Chief of the Weather Bureau who had succeeded Harrington in 1895, Congress was giving the Bureau "every dollar" it asked for to expand its public services to match the Nation's needs.

In 1908 Mr. Ford introduced his Model "T" and the country left the livingroom for the open road. Naturally, automobilists were concerned about the weather, and the Bureau's forecasts and warnings soon became a part of this facet of American life. As far as we know only one American motorist has ever been fined for not believing a weather forecast.

In 1916 a Kansas City man arrested for speeding pleaded he was on an errand of mercy. The weather was mild when his wife and daughter left for the theater and he had advised them not to wear heavy wraps, despite the published forecast of a sharp drop in temperature. A cold wave hit before the performance ended and he was racing to the theater, wraps in hand, when arrested. In imposing the usual fine the court said that since he had read the official forecast and ignored it, his excuse was inadmissable.

When the Weather Bureau took over from the Signal Corps in 1891, weather services were provided from the central office in Washington, D.C., by only four forecasters. Three years later the number had risen to 40 and district forecast centers had been established. Even with the district centers, however, forecasts were still prepared in Washington. A local forecaster could modify them to accommodate local conditions, but if he was wrong too often, he could lose his job. For under Chief Moore each forecaster competed with every other one in the country, and "survival of the fittest" was the rule.

The Signal Corps began 24-hour predictions in October 1872, and forecasts for 36 hours by July 1888. In 1898 the Weather Bureau extended this to 48 hours for forecasts based on the evening observations. Forecasting procedures followed by the Weather Bureau were similar to those of the Signal Corps, and continued with little change until the late 1930's when the Bureau began issuing four daily forecasts instead of two.

The Weather Bureau started experimenting with very general weekly forecasts in 1908 and was issuing them regularly by 1910. This early long range forecasting effort was undertaken

Willis Luther Moore was Chief of the Weather Bureau from
1895 to 1913.   —ESSA

The Weather Bureau forecast office in Washington, D.C., in
1926.   —ESSA

to help farmers plan their agricultural activities. In 1940 the
weekly forecast was replaced by a more detailed 5-day forecast,
and in 1950 after several years of experiments, the Bureau began issuing 30-day outlooks.

The river and flood forecasting service was expanded in the
Bureau's early years, but forecasters had only primitive instruments and a very thin reporting network so service was largely
confined to larger, more sluggish rivers where reasonably accurate forecasts could be made by rule of thumb. During periods
of high water the river station offices were besieged early and

late. By 1913 there were 483 stations in the river and flood network (16 operated in cooperation with the U. S. Geological Survey), but much remained to be done.

From 1924 through 1937 floods annually caused an average of 90 deaths and more than $100 million property damage. The 1927 floods in the Mississippi Basin alone killed 313 people and property damage was estimated at over $335 million.

Despite these disasters, the river and flood service was cut back during the Great Depression. From 1937 to 1941, however, funds from the U.S. Army Engineers Corps and the New Deal's Works Progress Administration supplemented money appropriated for the Department of Agriculture and made it possible for the Weather Bureau to greatly expand and reorganize its river forecasting system. By 1941 the river and flood service, though far from perfect, was approaching the limits imposed by the state of the art.

When the Battleship *Maine* was sunk in Havana Harbor in 1898, the United States declared war on Spain. The hurricane season was fast approaching; fearful for the safety of American naval units soon to be operating in hurricane waters, Chief Moore went to see President McKinley and urged the establishment of hurricane warning stations in the West Indies. McKinley, declaring he was more afraid of a hurricane than the Spanish Navy, ordered Moore to organize a warning system immediately.

A central hurricane forecast center was first established in Kingston, Jamaica, then moved to Havana, and finally, in 1902, transferred to Washington. During these first few years, the number of stations in the hurricane warning network was multiplied many times over.

In 1900 the worst natural distaster in our country's history hit Galveston, Texas. The great Galveston hurricane killed more than 6,000 people and completely devastated the Gulf Coast city. The hero of this terrible storm was Isaac Monroe Cline, chief Weather Bureau forecaster for Galveston, who lost his wife in the catastrophe.

Four days before the storm struck, the Weather Bureau in Washington warned that a hurricane was crossing Cuba heading northward; 2 days later storm warnings were issued for Galveston. On the morning of September 8, 1900, however, rising

President William McKinley ordered the Weather Bureau to establish a hurricane warning network in the West Indies.
—THE LIBRARY OF CONGRESS

tides and winds convinced Cline that the hurricane itself would roar ashore right over his city.

Though only the central office in Washington had the authority, Cline took it upon himself to issue hurricane warnings. By telephone and telegraph he alerted the city, then hitched his horse to a two-wheeled sulky and raced along the beach from one end of town to the other, yelling to the fascinated crowds still watching the great waves crashing ashore to move to higher ground. By late afternoon all telegraph and telephone lines were down and water was already waist deep in some parts of the city. Cline and his three daughters were among the few survivors of the 50 people who sought shelter in his home. His wife was drowned.

Dr. Cline had a long and distinguished Weather Bureau career. During the Mississippi Basin floods in 1927 he and his assistant, Willard McDonald, were credited with early warnings that enabled engineers to save New Orleans and prevented great losses of life and property.

The first hurricane observations radioed from ships at sea were received by the Weather Bureau in 1905. Efforts were made to expand this program so observations would be received before the storms hit land. Ships' captains, however, naturally avoided hurricanes, and the Bureau received few reports. This problem was not satisfactorily solved until after the Second World War and the introduction of hurricane-hunter aircraft, storm-tracking radar, and weather satellites.

The disastrous Labor Day hurricane that hit the Florida Keys in 1935 brought an appropriation increase for the hurricane warning service, which was reorganized that year. Additional forecast centers were established in Jacksonville, New Orleans, and San Juan, and a teletype network linking U.S. coastal cities was installed.

In 1910, fires in northern Idaho and western Montana consumed some 3½ million acres of prime timber and killed 50 people. The town of Wallace, Idaho was partially destroyed, and practically its entire population evacuated by the Northern Pacific Railroad. This conflagration led to the establishment of fire control operations by the U.S. Forest Service, and to the beginning of the Weather Bureau's fire weather services.

Weather conditions are primarily responsible for the occurrence and behavior of forest fires. Bureau forecasters at Portland, Oregon and San Francisco began issuing fire weather forecasts in the spring of 1913, and the Fire Weather Service was officially established on April 10, 1916. In the late 1920's Leslie Gray, fire weather forecaster at San Francisco, installed a two-way radio on a pickup truck and began going to the fires personally, preparing weather forecasts and briefing fire control bosses on the spot. By 1936, Gray had four more fire weather mobile units in operation.

In 1926 Congress made its first appropriation for fire weather work, which to that time had been conducted on a very small scale. Since 1926, fire weather services have been considerably refined and extended to most of our national forests.

Great Guns! Cyclone, Blizzard, and Snow
In its infancy, the weakest link in the Bureau's chain of service was often communications—getting its forecasts and warnings to the public quickly and in an accurate, useful form. Consider the public reaction to the following headlines in a newspaper of the 1890's:

GREAT GUNS!

WIND WILL BLOW A GALE

CYCLONE, BLIZZARD, AND SNOW

PREDICTED.

The original forecast? "Storm center over Manitoba, moving east; south winds will probably shift to west by Saturday morning; rain turning to snow Saturday."

The Weather Bureau continued to rely on signal flags and whistles to announce the "coming weather," while searchlights, sirens, and even rockets and bombs were used by local communities to warn of storms or cold waves. By 1901 Bureau forecasts were daily disseminated by telegraph, telephone, and mail

Storm signals fly from a tower on the Potomac River south of
Washington, D.C. Note the wind vane on the top of the tower.
—ESSA

to about 80,000 recipients. Weather maps and bulletins were posted in hotels, stores, office buildings, post offices, railway stations, and even trolley cars, while railroad trains carried weather signals on the sides of their baggage cars. Meanwhile, the recently inaugurated rural free mail delivery system was bringing the forecasts to heretofore inaccessible farming communities, and telephone operators were reading them to their companies' subscribers. By 1904 some 60,000 farmers in Ohio alone were getting the daily weather forecasts from Washington by telephone within an hour of their issue, and by 1926 there were some 5,500,000 rural subscribers to this service.

An automatic telephone answering service was introduced in New York City in 1939, and by the fourth day 58,000 calls were being recorded daily. Nationwide, these large-volume systems currently handle about 250 million public calls a year.

The Weather Bureau started using radiotelegraphy to reach the farmer in 1914. In January of that year arrangements were made for the broadcasting of forecasts by a station operated by the University of North Dakota. Nine amateur operators then made local distribution of the forecasts they received. In the early 1920's developments in radiotelephony made it possible to broadcast directly to the farmer, and radio soon showed signs of outstripping every other method of reaching rural areas.

If the wireless was a blessing to the farmer, it was a life saver to the sailor. Radiotelegraphy revolutionized ocean weather services, making it possible to collect current observations from ships at sea and to broadcast back forecasts and warnings.

Responsibility for marine meteorological work was transferred to the Weather Bureau from the Navy's Hydrographic Office in 1904. This was a continuation of the international cooperative climatological program begun by Matthew Fontaine Maury a half-century earlier. In addition, the Weather Bureau had long issued twice-daily forecasts of wind and weather as far as the Grand Banks for vessels leaving North American ports, while storm warnings were furnished to captains in port and displayed by flags and lights on the coast. Before the wireless, however, once a ship dropped below the horizon, it was on its own.

The first wireless weather report received from a ship at sea came from the S.S. *New York* on December 3, 1905. The following year 50 vessels were transmitting to the Weather

Bureau, and by 1925 nearly 35,000 reports were being received annually. In 1902 the Marconi Company began to broadcast Weather Bureau forecasts to the Cunard Company, and in 1904 broadcasts of storm and hurricane warnings by Navy radio stations for ships at sea became a regular service.

The Weather Bureau was initially built on the telegraph, and was long dependent on the free cooperation of the telegraph companies. In the early days of meteorology, many telegraphers were volunteer weather observers, working in turn for Espy, Henry, Abbe, the Signal Service, and finally, the Weather Bureau. Through all these years, weather messages got top priority —even when there was only a single wire open.

In 1900, the completion of an ocean cable from Lisbon and the Azores to New York made possible the exchange of weather observations and warnings with Europe, and by 1901 the Weather Bureau was routinely collecting simultaneous weather observations by telegraph and cable from an area comprising one quarter of the globe (by 1910 "practically the entire Northern Hemisphere"). The Bureau's dependence on the telegraph for collecting domestic weather observations ended in 1928, when a teletypewriter communications system was established to meet the growing demands of commercial aviation. By 1938 this system covered all but three of the 48 States.

The Air Commerce Age
Prior to World War I, there were few requests for flying weather forecasts. The pilot was his own weatherman; if he saw a storm coming he got out of the way. Except for the beginnings of a forecasting system in Germany for the Zeppelin airships, there was little direct application of meteorology to aeronautics anywhere in the world.

During the war, however, each major power built up an aviation weather service that was remarkably effective considering the haste with which it was organized and the general ignorance of weather conditions at flying altitudes. The big lesson learned was that pilots needed much more intensive and detailed services than the farmer, the seaman, or the average citizen.

Late in 1918 the Weather Bureau began issuing special bulletins and forecasts for domestic military training flights and the Post Office's new air mail routes between Washington and New York and Chicago and New York. In August 1919, daily flying weather forecasts covering the Nation were initiated for the Army and Post Office, and also released to newspapers in areas with substantial local flying.

In July 1919 the Weather Bureau's assistance to the British dirigible R34 during its round trip flight between England and Long Island received high praise from the British Air Ministry, particularly for the return trip. A rush message to the ship's commander got the dirigible safely off at midnight, ahead of an approaching storm and with exceptionally favorable wind and weather conditions.

Other transatlantic attempts, cross-country flights, and free-balloon races clamored for weather information. By 1920 there was a transcontinental mail route, and in 1924 anyone could leave New York one morning and be in San Fracisco the following evening. The Weather Bureau established its first flight forecast centers at Washington, D.C., Chicago, and San Francisco in 1920. Meanwhile, the War and Navy Departments were enlarging their own weather units, and Army and Navy meteorologists were being assigned to the Bureau's Washington forecast center to help provide special weather services to military pilots.

For aviation, the 1920's and 30's were decades of glamor, accelerating technology, and—most of all—personalities. Lindbergh made his lonely flight to Paris in 1927. Wiley Post and the *Winnie Mae* took the round-the-world record away from the *Graf Zeppelin* in 1931, and by the end of that year almost 500 people had flown the Atlantic. Lindbergh, Post, and most of the others had Weather Bureau briefings and forecasts for their flights.

When the 1930's began, pilots had already pushed super-charged biplanes beyond 40,000 feet and into the frozen stratosphere. Gradually, they would free themselves from the constraints of frail aircraft, rough runways, and poor communications—of everything, in fact, except the weather.

The Air Commerce Act of 1926 was the landmark legislation for commercial aviation. In the years to follow this new industry would grow at a fantastic pace. And the Weather Bureau, charged with providing meteorological support, would grow with it.

To accomplish its new mission, the Bureau established special observation stations along and near the airways, stations which reported the weather on an hourly basis, and on a minute-to-minute basis during rapidly changing conditions. Next, airport weather stations were established so the pilot could get personal briefings and forecasts before he took off. There were 50 airport stations by June 30, 1930, with some 250 airways sites providing weather observations. By 1941 the numbers had grown to nearly 100 airport stations and 823 airway reporting stations. Because of Weather Bureau budgetary limitations, the observers at airways stations were usually either military personnel or employees of the Bureau of Air Commerce (a forerunner of the Federal Aviation Agency).

As World War II approached, commercial air traffic became so heavy it required ground control. By the war's end the Civil Aeronautics Administration (successor to the Bureau of Air Commerce) had established 26 Air Route Traffic Control Centers, and each was supported by a Weather Bureau Flight Advisory Weather Services center.

Radio broadcasting of the Bureau's aviation forecasts began in 1921. By the early 1930's weather data and forecasts supplied to Bureau of Air Commerce personnel were being broadcast by airport control towers every 30 minutes.

A high altitude service for commercial aviation was started in 1959, with forecasts routinely issued for one third of the Northern Hemisphere. Today, weather services for aviation include automatic telephone answering systems, nationwide continuous weather broadcasts, and even closed circuit television briefings at busy international airports.

The basic contents of the Air Commerce Act and the later Civil Aeronautics Act were included in the Federal Aviation Act of 1958, which created the Federal Aviation Agency (FAA). The FAA and its predecessors (beginning with the Department of Commerce's Aeronautics Branch in 1926), as well as the military services, have been major working partners in the Bureau's efforts to provide weather services to aviation. In 1961, under a joint program with the Weather Bureau, 4,000 FAA specialists were trained to give weather briefings to pilots, making it possible to provide such services at hundreds of additional airports.

The importance and growth of the Weather Bureau's services to aviation brought about its transfer to the Department of Commerce on June 30, 1940, ". . . to permit better coordination of Government activities relating to aviation and to commerce generally, without . . . lessening the Bureau's contribution to agriculture." This transfer, after half a century in the Department of Agriculture, mirrored the transition in American life from a rural society to one approaching the Jet Age.

Up Where the Weather Is

For man, and particularly the weatherman, the atmosphere is an ocean above us. To forecast what will happen at the bottom, we have to know what's going on over our heads, up where the weather is. It was the air age which finally made it possible—and necessary—to systematically explore this airy realm, but the need had been recognized long before.

When the Signal Service was established in 1870, meteorologists had only surface observations to use in their attempts to understand and forecast the behavior of an active atmosphere many miles high. Then, in December 1871, Professor S. A. King, aeronaut, offered his services and his balloon.

Through the enthusiastic cooperation of Prof. King and other balloonists the Signal Service was able to send observers and instruments on aerial voyages of exploration. Cleveland Abbe went up on two early flights, after obtaining permission from his insurance company, and Professor Henry Allen Hazen (no relation to General Hazen) completed four ascents in 1885, followed by several more in 1886 and 1890, and even suggested a transatlantic flight.

Man's meteorological exploration of the atmosphere had, of course, begun long before the Signal Service. And it began with kites. In 1749 Professor Alexander Wilson of the University of Glasgow attached thermometers to the tails of several of a chain of kites and flew them. Protected by paper tassels, the thermometers fell to the ground unbroken after an ignited line disengaged them. Although Professor Wilson sounded the atmosphere to only a few hundred feet, he was the first to obtain observations from the free upper air, and established the kite as a meteorological vehicle.

A 1900 kite sounding at Arlington, Virginia, just outside Washington, D.C. —ESSA

For over 30 years, the kite remained the only means of sounding the ocean atmosphere. Then in 1783 Jean Francois Pilatre de Rozier of France made the first free balloon ascent, and weathermen soon began to accompany théir instruments aloft —often to perilous heights.

Although the information gained by the Signal Corps in its manned balloon ascents was extremely interesting, it was also fragmentary, infrequent, and often inaccurate. A better method was needed. Cleveland Abbe suggested kites, and following the International Conference on Aerial Navigation in Chicago in 1893, the kite was rapidly adopted by the meteorologists of many nations.

A kite reelhouse. White louvered box on the right is a meteorological instrument shelter.   —ESSA

In the United States, weather instruments were sent up in box kites in 1894 by Abbott Lawrence Rotch at Harvard's Blue Hill Meteorological Observatory, which he had founded in 1884. Professor Charles F. Marvin of the Weather Bureau, a future chief, began experimenting with kites in 1895, and a year later perfected his kite meteorograph, an instrument capable of measuring and ·recording temperature, pressure, and humidity. In 1898, the Weather Bureau opened 16 kite stations and attempted daily, simultaneous observations from April to November. Although light winds sometimes prevented flights, considerable information was obtained concerning the temperature profile of the lower atmosphere over a wide area of the central United States.

Bureau kite flying was shifted to Mount Weather, Virginia, site of the Weather Bureau's new research center (1,725 feet above sea level) in 1904. In 1907 one flight reached 23,111 feet, for which the Bureau claimed a world record. In that same year, a regularly scheduled program of kite observations was begun. Though it never involved more than seven stations, this programed probing of the atmosphere continued for a quarter of a century.

The flights were seldom trouble free. Piano wire was used and observations consisted of launching the kites, leaving them up for 4 or 5 hours, then reeling them back to retrieve the meteorographic record. The following account by a Bureau kite flyer in 1898 gives some idea of the problems that could arise.

> There seemed, on May 31, 1898, to be a thunderstorm approaching. . . . My reeler and I commenced to pull in the kite. When the kite was still 3,000 feet up . . . a bolt from the clouds was seen, by others, to hit the kite. We stopped reeling; we had to. I have been nearly in front of a 6-inch naval rifle at its discharge, and its noise was very mild compared to the report which we heard at the reel. . . . Fortunately, the kite reeler was rearing rubber overshoes, but I was not, and in consequence got a burn on the sole of my foot. . . . Several really ludicrous events have occurred when the kite has behaved badly. Sometimes the operators get knocked down and dragged about by the kite in attempting to land it, while the urchins, at a distance, shout humiliating taunts.

Because a wind of 10 to 15 m.p.h. was needed to raise the kites, it was difficult to obtain continuous, simultaneous observations. Even so, the kite was more practical than the manned balloon, and its use continued until 1933, when it was replaced by the airplane.

The rapid development of the airplane during World War I presented a novel and easy way to obtain upper air observations, and in 1917-18 Marvin meteorographs were mounted on the wings of military aircraft flying over France. The Weather Bureau began experimenting with its own aircraft observations in 1919, and 6 years later with the cooperation of the Navy's Bureau of Aeronautics, began daily observations at Washington,

Navy plane with aerometeorograph attached to strut between upper and lower wings. Picture taken at 9,000 feet over Washington, D.C. on December 13, 1934.
—U.S. Navy

Aerometeorograph chart from 1933 flight over Cleveland, Ohio
showing pressure, humidity, and temperature traces. —ESSA

Meteorologist preparing an atmospheric profile from aero-meteorograph traces. An aerometeorograph is shown in the background with the protective case removed.   —ESSA

D.C. Airplane soundings were made frequently at most naval air stations during the 1920's.

On July 1, 1931 the Weather Bureau began a regular program of early morning airplane observations at Chicago, Cleveland, Dallas, and Omaha, and 2 years later it closed its last kite station at Ellendale, North Dakota. By 1934 over 20 airplane soundings were being made daily by the Weather Bureau, Army, Navy, and Massachusetts Institute of Technology. The number increased to about 30 in 1937.

Growing Pains   57

Pilots, under contract to the Weather Bureau were paid for each flight, provided they reached an altitude of 13,500 feet (16,500 feet in 1938). A 10 percent bonus was paid for every 1,000 feet above the specified altitude.

Launching a score of aircraft across the country simultaneously was an expensive and dangerous way to get weather information. Between 1931 and 1938, 12 pilots were killed. In addition, it was often impossible to use the planes in bad weather, when their observations were most needed. For these reasons, and because of significant advances in the use of unmanned balloons to sound the atmosphere, airplane observations were discontinued before the Second World War.

Spurred by the needs of aviation, the Weather Bureau in 1909 began a regular program of free unmanned balloon observations which continues today. These small "pilot" balloons carry no instruments but are tracked with a surveyor's telescope to determine the wind direction and speed at various heights estimated from the balloon's known ascension rate. Since pilot balloons are tracked visually, they are practical only in clear weather or below clouds.

The Weather Bureau began attaching meteorographs and parachutes to its pilot balloons before World War I, to convert them to atmospheric sounding vehicles. When the balloon burst (due to the expansion of the hydrogen in the thinner air aloft) the instruments would float down safely by parachute, and their recorder traces could be recovered.

In 1927, two Frenchmen attached a radio transmitter to a balloon and established direct communication with the atmosphere. Shortly afterwards the Russian meteorologist, P. A. Moltchanoff, developed and balloon-flew a practical combination of radio and meteorograph. This was the ancestor of the present day "radiosonde," an instrument that measures temperature, humidity, and pressure and transmits their values at regular intervals.

The Weather Bureau began official radiosonde observations in 1937, and daily soundings a year later. In 1939-40 all Weather Bureau and military airplane stations were converted to the new sounding techniques. Procedures for radio direction finder (and later radar) tracking of the radiosonde ascents were developed

An early radiometeorograph sounding. Note the two balloons.
—ESSA

A "bedsprings" antenna identified the SCR-658 radio direction finder, used to track radiosonde soundings to obtain wind speed and direction. Developed by the Army and introduced in 1942, it was widely used by the military during World War II and adopted by the Weather Bureau about 1946. It was eventually replaced by lighter equipment. —ESSA

during World War II, making it possible to obtain wind direction and velocity measurements even after the balloon passes out of visual range or disappears into a cloud.

The radiosonde has given meteorologists a free flying, inexpensive way to profile the atmosphere to great heights. Two balloon sizes are used. The smaller, 600-gram balloons reach heights up to 90,000 feet; the larger, 1,200-gram balloons average more than 100,000 feet, with some exceeding 125,000 feet, or almost 24 miles.

The development of the radiosonde was an epoch landmark in meteorological history, leading to many revisions of man's concept of the structure of the atmosphere under, and in which, he lives. It is also aided in the development of new methods of weather forecasting.

Winds of Change

Dr. Charles F. Marvin, who had succeeded Willis Moore as Chief of the Weather Bureau in 1913, retired in 1934 after 50 years of Government service. He became Chief in 1913 and was head of the Bureau's instrument division for about 23 years before that, inventing or perfecting many of the Bureau's meteorological instruments. Marvin's retirement came just before the Bureau's adoption of the revolutionary work of Norwegian meteorologists during and after World War I, work which was to provide the vertical dimension meteorology needed to meet the accelerating needs of aviation.

During the First World War, the broadcast of weather observations and forecasts was banned by both sides to avoid giving useful information to the enemy. Norwegian scientists, cut off from the weather information they needed, were forced to seek new methods of forecasting. They focused their attention on the fact that most "weather" occurs at the boundaries between great air currents of differing temperatures and humidity. The continuous conflict at these boundaries resembled nothing so much as the battle then raging along the western front, and the name "front" was applied to the zones between different air currents or "airmasses." This concept led to the polar front theory and the airmass method of weather analysis. It provided an atmospheric model which meteorologists could use in explaining and forecasting the weather.

Professor Vilhelm Bjerknes and his son Jacob pioneered this revolutionary concentration on fronts and airmasses—especially the irregular outbreaks of enormous cold airmasses from the polar regions—rather than on individual storms, which represent only the interplay of airmasses. Their theory reshaped the science of meteorology.

Under Willis Ray Gregg, who succeeded Marvin, the Weather Bureau in 1934 established an airmass analysis section staffed by men trained in the new theory. The familiar fronts that appear on today's weather maps made their debut in 1936, and in 1938 Weather Bureau use of airmass analysis techniques

Francis W. Reichelderfer, Chief of the United States Weather
Bureau from 1939 through 1963.   —ESSA

62   Growing Pains

became official. Elsewhere, the ideas of the Norwegian school of meteorology had taken root much earlier.

Gregg, who had joined the Bureau in 1904, had been with the aerological (aviation) division since 1914, and had organized the Weather Bureau's services to aviation. When he died suddenly in 1938, he was succeeded by Francis Wilton Reichelderfer, who was appointed Chief on January 2, 1939. Reichelderfer was another pioneer in the development of weather services for aviation, and had introduced airmass analysis and forecasting techniques throughout the Naval meteorological service in 1925.

Gregg and Reichelderfer had been briefing partners for the Navy's historic NC-4 crossing of the Atlantic in 1919. Gregg was the Weather Bureau's meteorologist at Trepassey, Newfoundland, where the transoceanic flight began; Reichelderfer then a naval officer, the meteorologist at Lisbon, Portugal, the European terminal of the flight.

Reichelderfer was a Navy pilot, qualified for balloons, dirigibles, and airplanes. His introduction to the new theories came just after World War I, when he was assigned to the Naval Air Station at Hampton Roads, Virginia and became involved in aerial bombing exercises with an ex-German battleship. On one particular flight the forecast called for fair weather, but on their way home the planes ran into a severe line squall—what we would call a cold front today. Reichelderfer was in a multiengined bomber heavy enough to fly through. Billy Mitchell, however, along as an observer, was in a light single-engined plane and had to race southward just ahead of the northeast-southwest front to reach land and get down safely.

A short time after this incident, Reichelderfer came across the first of Bjerknes' papers concerning the theory of fronts and airmasses. He realized that this was what the Navy was looking for. From 1922 to 1928 Reichelderfer supervised the reorganization of the Navy's weather service, and its adoption of the Norwegians' new meteorological methods.

In 1938, Reichelderfer was promoted to Commander, then later transferred to the inactive list to accept appointment as Chief of the Weather Bureau. Under his administration the Bureau completed its transition to three-dimensional meteorology.

The NC-4 lands at Lisbon on May 29, 1919, after completing the first transatlantic flight to Europe (via Newfoundland and the Azores).
—THE SMITHSONIAN INSTITUTION

William Ferrell, a 19th century pioneer in theoretical meteorology.   —ESSA

Orville Wright is at the controls as the Wright brothers' plane lifts off the launching track on its historic first flight.
—The Smithsonian Institution

President William Howard Taft presents Aero Club of America
Gold Medals to Wilbur and Orville Wright on June 10, 1910.
—U.S. AIR FORCE

The Air Commerce Act of 1926 contained the first clearcut, formal authority for Weather Bureau research—research in support of aviation. This was followed by Public Law 657, which authorized the study of ". . . thunderstorms, hurricanes, cyclones, and other severe atmospheric disturbances. . . ."

Early weather service scientists like Cleveland Abbe pushed research as far as they could with the meager resources available, knowing that until basic knowledge of atmospheric processes was obtained, man could not do much more than follow rule-of-thumb forecasting methods and hope for the best. Abbe himself authored some 300 scientific papers and wrote thousands of letters encouraging other scientists to contribute to man's knowledge of the atmosphere and weather. In 1873 he started the *Monthly Weather Review* (still being published) which he edited until just before his death in 1916.

Dr. Carl Gustaf Rossby with a rotating tank constructed in 1926-27 for his early studies of atmospheric motion. —ESSA

Other pioneer meteorologists made even greater contributions. Toward the end of a long and distinguished career, William Ferrel worked for the Signal Service during the period 1882-86. Earlier he had successfully explained much of the mechanics of the atmosphere and of storms, and had pioneered a new field of "dynamic" or theoretical meteorology, based solidy on mathematics and physics.

Chiefs Harrington, Moore, and Marvin, each promoted research in turn, though none had specific appropriations. (Professor Marvin's own experiments with kites provided invaluable wind data to the Wright Brothers and other aviation pioneers, some of whom also corresponded with Cleveland Abbe.) All in all, however, there was little money for much more than individual efforts until the Air Commerce Act, which led to the eventual establishment of the Bureau's Meteorological Research Division in 1936.

One man who contributed greatly to the present state of meteorology was Carl-Gustaf Arvid Rossby, perhaps the most brilliant theoretical meteorologist of the 20th century. Rossby, a disciple of Vilhelm Bjerknes, came to the United States in 1926. After working for the Weather Bureau for 18 months he went to California on a Guggenheim grant to set up a model airway weather service. The system he established in collaboration with the Weather Bureau became the prototype for the Bureau's entire airways meteorological service. In 1928 Rossby was appointed head of the meteorology department of the Massachusetts Institute of Technology.

At MIT, Rossby and Hurd Willett began a major attack on the secrets of the atmosphere and its circulation. Their work coincided with the great Dust Bowl drought of the 1930's, and Rossby soon got financial support from the Department of Agriculture, which hoped that droughts and other weather disasters could be forecast.

The general atmospheric circulation causes and controls all weather and climate. In the 19th century it was known that the circular storms at the surface become wave-like ripples in the great westerly winds aloft over the Northern Hemisphere. In 1937, Jacob Bjerknes showed that these waves were hemispheric in scale and circled the globe. Rossby and his MIT colleagues discovered that the waves (now called Rossby waves) extend vertically through the greater part of the atmosphere and evolved a mathematical equation to predict their motion. Since these waves steer the cold and warm airmasses that cause surface weather, Rossby's work made it theoretically possible to make relatively long range forecasts for the Northern Hemisphere.

The first systematic attempt to use Rossby's circulation model for such forecasting was made by Hurd Willett, Jerome Namias (now Chief of the Weather Bureau's Extended Forecast Division), and their MIT coworkers in the fall of 1940 when they issued the first 5-day forecast. In 1941 the project was moved to Washington, D.C., and became an operational and research activity of the Weather Bureau. Meteorology had turned an important corner. So had the United States . . . World War II was almost upon us.

Dust storm approaching Springfield, Colorado.
—DEPARTMENT OF AGRICULTURE

With war imminent, the military were very interested in long range weather forecasting. While 5-day forecasts were of great value, the need for forecasts covering much longer time periods led to research to produce a 30-day prediction. This effort began in 1942. In 1945 experimental 30-day "outlooks" were prepared for selected defense industries. After the war the Weather Bureau began issuing the outlooks to the public, in much their present form.

These 30-day outlooks are not forecasts in the usual sense, but rather give the temperature and precipitation trends expected for the coming months in terms of climatological normals for the period.

During the war, 5-day forecasts were prepared twice a week and distributed to the military weather services. Special forecasts were also prepared for maneuvers and military operations, and Namias and K. E. Smith of the extended forecast unit received Navy citations for their work in connection with the Allied invasion of North Africa.

Modern meteorology really came of age during the Second World War. For it soon became obvious that success in this war, more than in any previous war in history, would often depend on whose side the weather was on.

PART III
WEATHER IN WAR

Weather in War

In September 1939 Hitler's panzers raced unchecked across Poland over hard dry ground and under clear blue skies specked with the broken-wing silhouettes of Luftwaffe Stukas. Three years later the Russian winter broke the back of these same panzer divisions, leaving them helpless before the Red Army.

A bogged-down tank, a grounded airplane, an army cut off from its supplies has no striking power. Weather is a key to success in warfare. Only during the last century has it become possible to wield this weapon with any skill; only since World War I has the meteorologist become a major member of the military staff.

Warrior Weathermen

United States entry into World War I in 1917 created an immediate military need for accurate and detailed weather information for overseas operations. The inability of the domestically oriented Weather Bureau to meet these meteorological demands led to the growth of military weather services. When the war ended, these meteorological activities were retained, though drastically reduced in size, as an essential function of command.

The really significant commitment of America's Armed Forces to meteorology, however, came with World War II. As in the First World War, the military's need for meteorological services was beyond the resources of the Nation's civilian agency, so the military weather services had to fill the void.

Uninformed weathermen accompanied America's fighting forces to all theaters of war. One of them, Army Air Corps Captain Robert M. Losey, was the first U.S. officer killed by enemy action in World War II. Military meteorologists set up stations alongside advanced airstrips, and were with the first waves that stormed ashore on Pacific atolls. Five were killed during the Japanese attack on Pearl Harbor, and fifteen were last-ditch defenders of Bataan and Corregidor. They helped plan the invasions of Africa and Europe, and parachuted and waded onto Normandy's beaches to give on-the-spot support to the D-Day landings.

After World War II, the need for occupation forces and then the cold war made it necessary for the United States to retain strong armed services, and consequently strong weather support organizations. These meteorological services have played a major part in the development of meteorology in the United States. Their efforts have benefited the American public in peacetime, at least as much as the military in time of war.

In the Footsteps of Myer
In the First World War, the need for a weather service within the American Expeditionary Force (AEF) was evident even before the arrival of American troops in France and led to the creation of a meteorological section in the Science and Research Division of the Signal Corps. The new section was charged with providing the AEF—as well as Army aviation, artillery, ordinance, and gas warfare activities in the United States—with all the meteorological information it needed.

The Army's needs fell into two categories: weather forecasts for military operations and upper air soundings for aviators, balloonists, and artillerymen. To meet the first objective, E. H. Bowie of the Weather Bureau was commissioned a major in the Signal Officers' Reserve Corps and assigned to the staff of General Pershing as principal forecaster. Dr. William R. Blair, who had been in charge of the Bureau's aerological investigations, was also commissioned a major and given charge of Army aerological work. Major Blair organized the Signal Corps' weather observation station network in France.

The only existing source of trained meteorological personnel was the Weather Bureau, which soon found itself shorthanded to perform its own wartime mission. To solve the problem, inductees with the necessary educational background were trained as military weathermen. The first recruits received their instruction at Weather Bureau stations in the United States, then went to France for further training. In May 1918, a special school for

The 148th Aero Squadron on the flight line during World War I.
—THE NATIONAL ARCHIVES

Army plane equipped with aerometeorograph at Scott Field, Illinois in 1935.
—ESSA

stateside indoctrination was established at Texas A&M University, where Dr. Oliver L. Fassig and C. F. Brooks of the Weather Bureau were the chief instructors.

Men were sent overseas in groups of 50, until approximately 300 had gone. Others were sent to 37 Signal Corps stations at military training centers in the United States. Captain B. J. Sherry and Lieutenant Alan T. Waterman (later Director of the National Science Foundation) were responsible for the Corps' meteorological service in the States. During World War II, Sherry supervised the Meteorological Depot of Signal Corps, which supplied all the weather instruments and equipment used by the Army and the Army Air Forces in that war.

The first Signal Corps weather station in France was established in May 1918; the first station involved in combat operations supported the First Army Corps near Chateau Thierry. Altogether 15 observation stations and a complete meteorological headquarters and forecast center were established before the war's end. Each station furnished weather information to local military units and took the observations the headquarters station needed to prepare forecasts and general weather bulletins. Observations were exchanged with the British and French meteorological services to extend the area of coverage.

Weather information was an important element in both the planning and execution of military operations in World War I. The great German offensive that began on March 21, 1918 took advantage of dry weather that was likely to last for several days. The pressure pattern also promised favorable winds for the use of gas. This was critically important; a few weeks later four Prussian regiments had their gas blown back in their own faces following a sudden windshift during an attack on Armentieres.

When the war ended, the meteorological activities of the Signal Corps were sharply reduced. The growth in postwar military aviation, however, created a demand for aerological information and services, and a Signal Corps weather officer was soon assigned to all major Air Corps maneuvers. The Signal Corps' Meteorological Section also contributed to the success of such important events as the first round-the-world flight, made by Army aviators in 1924.

In 1937, responsibility for meteorological support was trans-

80   Weather in War

Start of an Army pilot balloon run in the 1930's. The instrument on the tripod (called a theodolite) is used to track the balloon as it rises, to obtain a profile of wind direction and speed aloft. The observer is connected by telephone directly to a data plotter (b) in the weather shack.   —U.S. AIR FORCE

ferred from the Signal Corps to the Army's various operating services—the Air Corps, the Artillery, and the Chemical Corps. The Signal Corps continued to be responsible for the Army's meteorological research and for the development and procurement of meteorological instruments, equipment, and supplies.

Growth of a Global Weather Service
In 1937 the United States ranked sixth among the nations of the world in combat air strength. Early German successes in Europe, however, quickly demonstrated the value of air power in modern warfare, and led to the development of a U.S. global air strike capability—and of our largest military meteorological organization, the Air Weather Service.

Forty weather stations manned by 22 officers and 180 enlisted men were transferred from the Signal Corps to the Army Air Corps on July 1, 1937 to support aviation; this was the approximate size of the Air Corps' weather service 2 years later, when war came to Europe. In the summer of 1944, when the Service reached peak strength, there were 19,000 officers and men manning some 900 stations. Nearly 700 of these stations were overseas, many in primitive areas where no weather service had ever existed.

Two of the officers most intimately involved in the development of the new Air Corps weather service were Captain Randolph P. Williams, the original organizer of the service, and his successor, Captain Robert M. Losey. Both were killed in the war. Captain Losey died in a German air raid on the town of Dombas, Norway in 1940. Captain Williams was killed in 1944 when the photo reconnaissance plane in which he was riding was shot down by enemy aircraft over Saint-Michael, France.

The first weather reconnaissance squadron was activated in August 1942. The following summer its planes were monitoring the aircraft ferry route to England, radioing back weather conditions to the many young and inexperienced crews waiting to make their first ocean flights.

Reconnaissance flights were also the only practical way to determine weather conditions over the European Continent. These weather scouts flew their fighter aircraft to the primary and secondary targets, recorded winds, cloud cover, and cloud heights, then radioed this information back to the bomber strike

The Weather tower at the Murmitola Army Air Base in India
in 1943.   —U.S. Air Force

An Air Force meteorologist making weather and cloud calculations for a B-17 bombing mission. Bassingbourne, England, 1943.
—U.S. Air Force

force. Although their primary job was to be the meteorological eyes of the bombers, they frequently went to the defense of the formation when it was attacked by enemy aircraft.

In the fall of 1942 under the direction of General Hap Arnold, the Army Air Forces (renamed in 1941) began providing observations and forecasts to other Army Forces, which usually were not equipped to provide the extensive weather support they frequently needed. By 1943, the 12th Weather Squadron was furnishing meteorological information to both air and ground forces in North Africa and Italy, while in England Air Force meteorologists were making plans to support the tactical air and ground forces committed to the forthcoming invasion of Europe. In May 1945 the Army Air Forces' Weather Service was asked to furnish weather forecasting services for all Army components except those that might be specifically exempted by the War Department.

Air Force weathermen were stationed at many isolated outposts during the war. One was the only member of the U.S. Armed Forces on St. Helena—Napoleon's island of exile. Others were sent by dog team and tractor to the top of the Greenland Ice Cap. The most dangerous assignment, however, was to take and transmit weather observations behind enemy lines.

In the late winter of 1943 weather task teams parachuted into the mountains of Yugoslavia to support a planned bomber strike on the oil refineries of Ploesti, Romania. After lugging their equipment through deep snows, the teams set up observation posts in partisan camps. They ran their radios with handcharged batteries and dodged strafing attacks, artillery fire, and German patrols. The weathermen stayed in Yugoslavia until autumn, after the strike had been successfully carried out.

The Ploesti raid is a good example of the wartime relationship between meteorology and military planning. When the Army Air Forces decided to launch an all-out attack on the Romanian oil center, their nearest bases were in North Africa. This meant operating at nearly the maximum range of the B-24, the longest range bomber available, with the greater part of the route over heavily armed enemy territory.

There could be no strong headwinds, either coming or going, or the planes would never return. Since no fighters had enough

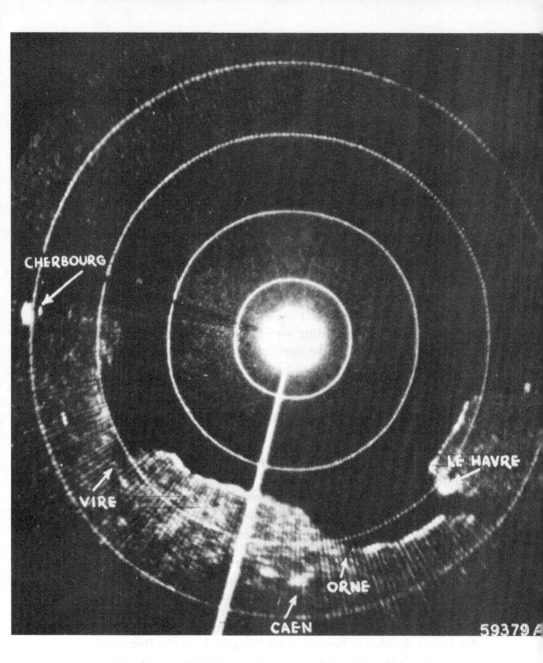

The Normandy beaches as they appeared on the radarscope of
Air Force bombers during pre-invasion reconnaissance flights.
—U.S. Air Force

86   Weather in War

Tempelhof Air Force Base in Berlin on March 1, 1949, following a snowstorm. Later that night high winds, poor visibility, and icing conditions forced a temporary halt to Airlift operations. —U.S. AIR FORCE

Mobile meteorological stations carrying Air Force weather personnel moved inland with the invasion troops.
—U.S. AIR FORCE

German cylinder chimney radar installation on Normandy
Beach after the D-Day invasion.   —U.S. Air Force

range to escort the bombers, protective clou
en route. Over the target itself, the bombers
and southerly winds—the latter so the plar
incendiary bombs on approach and let the w
rather than having to make additional pas
defended oil refineries.

By studying daily weather maps for th
Air Force weathermen found that climatolo
ditions were almost a sure thing in March o
occurred in any other month. March was
was selected.

Forecasters at the North African bases
for the right weather situation. As luck
weather was right on August 1. Though l
raid was a success. It might easily have b
for the careful climatological planning an
that preceded it.

Probably the most critical and comple
World War II was the Normandy invasi
prolonged period of high winds to produce
hamper the landing craft. Pilots wante
troopers cloudy skies to protect them
The Army, fearing a gas attack, want
Navy wanted small waves, thus offshore
landings had to be made at low tide, ar
least three good weather days to bring
ashore after D-Day. If the weather con
invasion might have to be put off for m

A joint meteorological staff had been
sisting of weathermen from the British
alty and from the U.S. Air Forces wea
groups were quartered separately so
could not wipe out the entire weather
ditions that seemed best for all invol
asked the meteorologists to calculate t
of getting the desired weather. The od
for May; 13 to 1 for June; and 50 to
tentatively chosen.

By June 3 it was obvious the wea
enough, so Eisenhower postponed the

the evening of the 4th, the weathermen predicted relatively good weather for the 6th, and the signal was given to go. The next day, the last chance for cancellation, the American team still said go, one British team said no. The other British group was undecided; when it joined the Americans, Eisenhower was briefed and the invasion was on.

D-Day caught the Germans by surprise. Later Major Lettau, the chief German meteorologist, said he had told the German high command an invasion would be impossible in the days immediately following June 4 because of bad weather. Had the invasion been postponed until the next time moon and tide conditions were right—June 17-21—it would have encountered the worst June storm in the English Channel in 20 years.

When the fighting ended the Army Air Forces meteorological service undertook the rehabilitation of the national weather services of many war-ruined countries. It transferred stations and equipment to local control and trained meteorological personnel. Italian, French, Dutch, Yugoslavian, Greek, Danish, Icelandic, Belgian, Indian, Chinese, Brazilian, Filipino, Korean, and German weathermen were among those trained by AAF meteorologists.

Following the war, the Air Forces' weather service was made a global command and in 1946 redesignated the Air Weather Service. To get observations from areas where it is impractical or impossible to operate ground stations, the Service initiated regularly scheduled weather reconnaissance flights from Guam, Japan, Hawaii, Alaska, Puerto Rico, and California, flights which still log millions of air miles annually.

The Air Weather Service had not yet recovered from the rapid post-war demobilization which reduced its strength from 19,000 to 9,000 men when the 1948 Berlin Airlift began. Weather was one of the greatest problems that American and British airmen faced, with low clouds, fog, freezing rain, turbulence, and ice of critical concern. The Air Weather Service met the challenge by assigning its most experienced forecasters to the operation and introducing such innovative techniques as telephone conferences between scattered airlift meteorologists to produce composite forecasts for the terminals and flight corridor.

Captain Willis J. Gran, weather officer at Da Nang Air Base,
South Vietnam, 1966.   —U.S. AIR FORCE

Mobile meteorological stations carrying Air Force weather personnel moved inland with the invasion troops.

—U.S. Air Force

German cylinder chimney radar installation on Normandy
Beach after the D-Day invasion.   —U.S. Air Force

range to escort the bombers, protective cloud cover was needed en route. Over the target itself, the bombers required clear skies and southerly winds—the latter so the planes could drop their incendiary bombs on approach and let the wind spread the fires, rather than having to make additional passes over the heavily defended oil refineries.

By studying daily weather maps for the previous 40 years, Air Force weathermen found that climatologically all these conditions were almost a sure thing in March or August, but seldom occurred in any other month. March was too soon, so August was selected.

Forecasters at the North African bases were alerted to watch for the right weather situation. As luck would have it, the weather was right on August 1. Though losses were heavy, the raid was a success. It might easily have been a total failure but for the careful climatological planning and weather forecasting that preceded it.

Probably the most critical and complex weather problem of World War II was the Normandy invasion. There could be no prolonged period of high winds to produce swell heavy enough to hamper the landing craft. Pilots wanted clear weather, paratroopers cloudy skies to protect them from German planes. The Army, fearing a gas attack, wanted onshore winds; the Navy wanted small waves, thus offshore winds. In addition, the landings had to be made at low tide, and the Allies needed at least three good weather days to bring equipment and supplies ashore after D-Day. If the weather conditions were wrong, the invasion might have to be put off for more than a year.

A joint meteorological staff had been established in April, consisting of weathermen from the British Air Ministry and Admiralty and from the U.S. Air Forces weather service. The three groups were quartered separately so a single German bomb could not wipe out the entire weather staff. After choosing conditions that seemed best for all involved, the high command asked the meteorologists to calculate the climatological chances of getting the desired weather. The odds *against* it were 24 to 1 for May; 13 to 1 for June; and 50 to 1 for July. June 5 was tentatively chosen.

By June 3 it was obvious the weather would not be good enough, so Eisenhower postponed the invasion until the 6th. On

Tempelhof Air Force Base in Berlin on March 1, 1949, following a snowstorm. Later that night high winds, poor visibility, and icing conditions forced a temporary halt to Airlift operations.
—U.S. Air Force

the evening of the 4th, the weathermen predicted relatively good weather for the 6th, and the signal was given to go. The next day, the last chance for cancellation, the American team still said go, one British team said no. The other British group was undecided; when it joined the Americans, Eisenhower was briefed and the invasion was on.

D-Day caught the Germans by surprise. Later Major Lettau, the chief German meteorologist, said he had told the German high command an invasion would be impossible in the days immediately following June 4 because of bad weather. Had the invasion been postponed until the next time moon and tide conditions were right—June 17-21—it would have encountered the worst June storm in the English Channel in 20 years.

When the fighting ended the Army Air Forces meteorological service undertook the rehabilitation of the national weather services of many war-ruined countries. It transferred stations and equipment to local control and trained meteorological personnel. Italian, French, Dutch, Yugoslavian, Greek, Danish, Icelandic, Belgian, Indian, Chinese, Brazilian, Filipino, Korean, and German weathermen were among those trained by AAF meteorologists.

Following the war, the Air Forces' weather service was made a global command and in 1946 redesignated the Air Weather Service. To get observations from areas where it is impractical or impossible to operate ground stations, the Service initiated regularly scheduled weather reconnaissance flights from Guam, Japan, Hawaii, Alaska, Puerto Rico, and California, flights which still log millions of air miles annually.

The Air Weather Service had not yet recovered from the rapid post-war demobilization which reduced its strength from 19,000 to 9,000 men when the 1948 Berlin Airlift began. Weather was one of the greatest problems that American and British airmen faced, with low clouds, fog, freezing rain, turbulence, and ice of critical concern. The Air Weather Service met the challenge by assigning its most experienced forecasters to the operation and introducing such innovative techniques as telephone conferences between scattered airlift meteorologists to produce composite forecasts for the terminals and flight corridor.

Captain Willis J. Gran, weather officer at Da Nang Air Base,
South Vietnam, 1966.   —U.S. AIR FORCE

In 1950 less than 24 hours after the fighting began, the first Air Weather Service reconnaissance aircraft was flying over Korea; just 48 hours after the enemy crossed the 38th Parallel, AWS observers and forecasters stationed at Taegu were sending weather information to United Nations forces. Throughout the conflict, Air Weather Service meteorologists continued to give close support to U.N. forces, while its reconnaissance aircraft made daily flights over the battlefields.

In January 1962 the Air Weather Service once again responded to the call for wartime meteorological support by sending 23 weathermen to Vietnam. By 1969 the number had grown to 600, and three weather squadrons were operating at more than 30 locations throughout South Vietnam and Thailand.

Air Force combat weather teams operate in the field with the Army and also support airborne units, parachuting into battle areas with them. In 1967 General William C. Westmoreland said, "No other U.S. military commander ever had the advantages of the outstanding weather support I have had at my disposal."

Today Air Weather Service Headquarters at Scott Air Force Base in Illinois directs the operations of a worldwide network of weather facilities which provide round-the-clock support to Air Force and Army units at all levels. More than 12,000 officers, airmen, and civilians man its 400 units, which include 3 major weather centrals (one in Nebraska, one outside London, and one in Tokyo), 8 forecast centers, 257 forecasting and observation stations, and scores of smaller units.

The Air Force Global Weather Central at Offutt Air Force Base, Nebraska, is the largest operational weather analysis and forecasting facility in the world. In this underground headquarters of the Strategic Air Command Headquarters, Air Weather Service personnel operate a fully automatic meteorological support system. Communications networks and high speed computers bring weather data from the overseas centrals into the Center at the rate of 4,500 teletype words a minute, and as many as 10 million weather elements are considered and checked by computer before being incorporated into machine-prepared weather analyses and forecasts tailored to the needs of military users. These products are immediately disseminated throughout the world by teletype and facsimile transmission.

Weather was often the decisive factor in the success or failure of naval operations in World War II. —U.S. COAST GUARD

The Navy's weather central and camp at Petropavlovsk, U.S.S.R., during World War II. The Navy also operated a weather central at Khabarovsk.   —U.S. NAVY

Today the Air Weather Service finds itself increasingly concerned with "weather" farther and farther from the surface of the earth. As newer missiles and more sophisticated space age vehicles and weapons are introduced, they require new and more accurate knowledge and forecasts of conditions in the earth's upper atmosphere and beyond. The Air Weather Service has already completed its first decade of space weather support.

War on the Water
Throughout World War II, sailors and pilots of every belligerent nation scanned the skies, seeking the moment of weather advantage over the enemy. The attack on Pearl Harbor came under cover of frontal cloudiness, allowing the Japanese fleet the element of surprise. Later, in the decisive battle of Midway, Japanese forces managed to get within 600 miles of the island without detection, again using a weather front to cover their approach. At the critical moment of attack, however, the front disintegrated, exposing the task force to American land- and carrier-based aircraft, and the Japanese Navy suffered its first decisive defeat in 350 years.

It was no different in the Atlantic. In February 1942 the Germans used squally, foggy weather to cover the escape of the battleships *Gneisenau* and *Scharnhorst* from Brest and screen their subsequent flight northward through the English Channel. Earlier, German U-boats taking advantage of similar weather and sea conditions in the North Atlantic almost succeeded in severing the lifeline to Europe.

During World War II, extensive naval operations in two great oceans made unprecedented demands on the Navy's relatively small Aerological Service. Pearl Harbor found the Service with 90 officers and 600 enlisted men. By August 1945, these numbers had grown to 1,318 officers and approximately 5,000 enlisted personnel.

The U.S. Naval Weather Service was born during the First World War, when it became evident that weather played a decisive role in the success or failure of aircraft operations and Zeppelin air raids. At the suggestion of the Commander of U.S. Naval Forces in Europe, the first step toward establishing a special meteorological organization to meet the needs of naval

The graduating class of the Navy's aerology school at Pensa-
cola, Florida, July 1923.   U.S. NAVY

The Navy's aerological office at Pearl Harbor was located on top of a water tower in 1925. Presumably the plane in the inset (top) carries an aerometeorograph. —U.S. NAVY

Radiosonde balloon release from the fantail of a Navy transport.
—SEALIFT MAGAZINE

aviation was taken in December 1917. Assistant Secretary of the Navy Franklin Delano Roosevelt asked Dr. Alexander McAdie, Director of Harvard University's Blue Hill Observatory, to join the Naval Reserve and organize a naval aerological service. An intensive course in meteorology for Naval officers was quickly inaugurated at Blue Hill, and a school for enlisted weather observers organized at Pelham Bay, N.Y. By the end of World War I, 50 officers and approximately 200 enlisted men were staffing naval weather support units at 26 stations in Europe and 5 in the United States.

In 1919, the Naval Aerological Service was established on a permanent basis in the Bureau of Navigation where, as then deemed appropriate, it was combined with the Photography and Pigeon Section. Two years later, it was transferred to the new Bureau of Aeronautics.

Demobilization reduced the number of trained personnel to approximately 8 officers and 12 enlisted men. With the subsequent growth of naval aviation, however, the Aerological Service was again expanded. By 1928 there were more than 100 officers and men manning observation and forecasting stations, and aerological units were assigned to the fleet flagship, flagships of divisions and squadrons, and aircraft carriers.

After World War II, a trend toward economy again reduced the size of the Service, but the Korean conflict in 1950 brought immediate expansion. The Navy's weather service has remained at approximately the same strength since, about 2,500 officers and men.

In recent years, accelerating advances in the electronic, supersonic, and nucleonic capability of the Navy had demanded more sophisticated weather support. On December 15, 1959, because of its broader mission, the Naval Aerological Service became the Naval Weather Service, and today serves the needs of the fleet and outlying outposts, as well as those of naval aviation.

The worldwide commitment of the Naval Weather Service is shown by the location of its weather activities. Besides those in the continental United States, Alaska, and Hawaii, there are Navy weather centrals or facilities in Iceland, England, Spain, the Philippines, Japan, the Antarctic, and on Guam. These centers—supported by a large network of secondary shore installations and small shipboard and staff weather units—provide general meteorological services to the Navy as well as oceanographic forecasts to all the Armed Services.

The DEEPFREEZE Weather Center at McMurdo Sound in the Antarctic operates at full tilt during that region's summer season (October through February). Hundreds of flights are cleared to U.S. and foreign stations annually, and flights along the 2,250-mile track between Christchurch, New Zealand and McMurdo are monitored from weather offices at each terminal.

Since 1961, the Aerial Ice Reconnaissance Unit of the Naval Oceanographic Office, successor to the Hydrographic Office, has surveyed ice conditions for shipping in Antarctica's Ross Sea (the Oceanographic Office began forecasting ice conditions for Antarctic waters in 1964). By following the sun, aerial ice reconnaissance airmen, known less formally as Polar Prowlers, are able to protect shipping at both the bottom and top of the world, completing their antarctic operations in time to begin arctic operations. Using their reports, the Naval Weather Service's Environmental Detachment at Argentia, Newfoundland, provides briefing and ship routing services for arctic waters between North America and Greenland and Iceland.

The Naval Weather Service's Optimum Track Ship Routing Program was inaugurated in 1958. Under this program, Navy meteorologists recommend sea routes where weather and ocean conditions are likely to be most favorable for a ship, task force, or fleet—whether to reach a destination quickly or carry out a military mission. Fleet Weather Central at Alameda, California serves the Pacific; Weather Central at Norfolk, Virginia, the Atlantic. Several thousand recommendations are made annually.

On July 1, 1967, in organizational recognition of its importance to the modern Navy, the Naval Weather Service was made a separate command.

Ocean Observations and Lifesaving

High Sea weather stations in the North Atlantic were suggested by Weather Bureau Chief Willis Moore before World War I. During World War II they became a reality.

With the onset of war in Europe in 1939, ships of belligerent nations maintained radio silence crossing the Atlantic. Soon after this, the Neutrality Act laid up American vessels in the European trade. As a result, practically no weather reports were

The first meeting of the Antarctic Chapter of the American
Meteorological Society. Except for Chapter President James
Beall of the Weather Bureau (bottom left) and the guest
speaker, Dr. E. T. Apfel (center) of the Department of the
Interior, all the members are Navy weathermen. The group is
in Antarctica to take part in Operation HIGH JUMP, a
scientific undertaking of the late 1940's.   —U.S. NAVY

available from the North Atlantic. The chance of war and the increasing number of American transoceanic flights made it imperative to obtain such reports; the practical way seemed to be to place ships "on station" at strategic spots in the Atlantic.

On January 25, 1940, President Roosevelt ordered Coast Guard cutters engaged in the neutrality patrol off Newfoundland to withdraw and man ocean weather stations. Two stations were established between Bermuda and the Azores, with the Coast Guard providing the ships and communications and the Weather Bureau, the meteorological personnel and equipment.

Great Britain suffered heavy shipping losses in 1940, and the transportation of lendlease aircraft to England soon became a critical problem. It was decided to fly American-made bombers directly from Newfoundland to England. This made a third ocean station vessel necessary, located about 500 miles northeast of Newfoundland.

When the war ended, there were 22 Atlantic stations, with 11 manned by the United States. (The British manned seven others, and Brazil supplied vessels for the remaining four.) Many of these ocean stations were established primarily for air-sea rescue and to provide the navigational (radio beacon) aid to pilots, with weather observations a secondary mission. One vessel, the U.S. Coast Guard Cutter MUSKEGET, was sunk by the Germans during the war, going down with all hands.

The U.S. Navy operated a similar ocean station vessel program in the Pacific during the war. It began in 1943 with one station in the Gulf of Alaska and another between California and the Hawaiian Islands. The program grew until January 1946, when there were 24 stations, including one manned by the Canadians. Currently, there are three Pacific stations, two manned by Coast Guard cutters with Weather Bureau meteorological personnel and one by Canadian vessels. When World War II ended, the Atlantic weather station program nearly ended with it. Thanks to the continuing growth of trans-Atlantic flight, however, steps were soon taken to put the program on a permanent basis.

On September 17, 1946, representatives of 11 nations met at an International Civil Aviation Organization conference in London to design a peacetime network of North Atlantic weather stations that would also provide navigation and air-sea rescue

The bow of the USCGC OWASCO after a winter storm in the North Atlantic during a 1962 patrol on ocean station "Bravo" (between Labrador and Greenland). Gusts of 75 and 80 m.p.h. picked up spray from the waves and deposited it as ice. Accumulations were up to 10 inches thick.
—U.S. COAST GUARD

One of the new class of Coast Guard cutters scheduled to replace the existing fleet. These 2,800-ton, 378-foot vessels are formidably equipped for meteorological and oceanographic work, as well as for air-sea rescue. An 80-foot flight deck (rear) gives them a helicopter search and rescue capability superior to that of any other ship smaller than a cruiser.

—U.S. COAST GUARD

A surfboat from the USCGC BIBB approaches the ditched Flying Boat BERMUDA SKY QUEEN to take off passengers and crew members.

—U.S. COAST GUARD

services. The conference resulted in an international agreement to establish 13 permanent ocean stations. The United States subsequently manned eight of these, but the number was reduced to the present four in 1954 when other Atlantic nations assumed a larger portion of the program. Great Britain, France, and Norway currently man five stations, while many nations contribute to the support of the program.

All U.S. stations are on major transoceanic air routes. When on station, Coast Guard cutters make regular weather broadcasts and furnish special meteorological assistance to passing planes and ships on request. They also provide checkpoints for aircraft navigators, and are prepared to come to the immediate aid of a ship or plane in distress. With one of these cutters standing by, the chances of ditching an airplane and surviving—even in storm-whipped seas—are very high. One of the most spectacular illustrations of this occurred in 1947, shortly after the peacetime network was established.

The flying boat *Bermuda Sky Queen*, carrying 62 passengers and 7 crewmen, exhausted most of its fuel bucking headwinds and could not make land. Despite 35-foot waves, the plane successfully landed near the U.S. Coast Guard Cutter BIBB, on station about 800 miles east of Newfoundland.

For 8 hours the cutter stood by, waiting for the seas to subside. Then the plane began to leak. Working against towering waves, BIBB crewmen made three round trips on a large, bobbing raft, ferrying passengers from plane to ship. On the fourth trip with 16 on board, the raft was swept away by the sea. A motor surfboat went to the rescue, but both raft and boat were swamped. Quickly the BIBB put men over the side in landing nets, and passengers and crewmen alike were plucked from the mountainous waters.

When darkness came, further rescue attempts were impossible, so 22 people spent the night on the storm-tossed seaplane. They were taken off the following morning, and 4 days later the BIBB steamed into Boston with all 69 survivors. If the BIBB had not been out there, it is doubtful there would have been any survivors.

The ocean station vessel program is not the Coast Guard's only weather-related mission. On April 14, 1912 the TITANIC went down with more than 1,500 people after striking an ice-

The luxury liner TITANIC departs Southampton, England prior to her maiden Atlantic crossing. —THE NATIONAL ARCHIVES

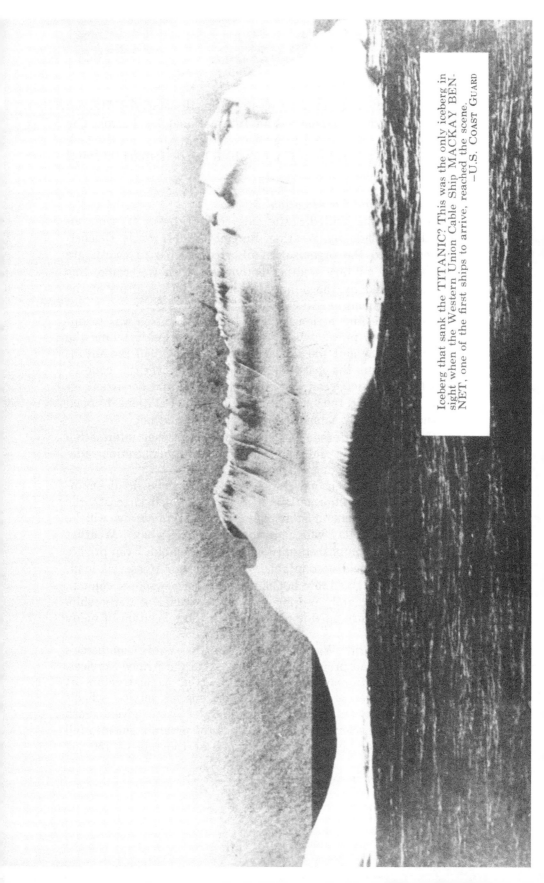

Iceberg that sank the TITANIC? This was the only iceberg in sight when the Western Union Cable Ship MACKAY BENNET, one of the first ships to arrive, reached the scene.
—U.S. Coast Guard

berg off Newfoundland. As a result of this disaster, the Atlantic maritime nations established an International Ice Patrol. The United States was asked to implement the patrol, and the task was assigned to the Coast Guard, which has performed this vital service ever since.

Cooperation and Coordination

Even before the United States entered World War II, a recommendation made by Weather Bureau Chief Francis Reichelderfer led to the organization of a committee to coordinate civilian and military weather activities. When war came, this was succeeded by the Joint Meteorological Committee of the U.S. Joint Chiefs of Staff.

One of the first actions of the Joint Committee was to approve plans to prevent weather information from reaching the enemy. Throughout the war, the United States had the advantage of knowing the eastward-moving weather over this country and the Atlantic Ocean. Despite determined and persistent efforts by spies and the use of automatic weather stations dropped in the Atlantic by submarine, Hitler usually did not.

The Weather Bureau began coding all weather information broadcast by Navy and Civil Aeronautics Administration radio stations on the evening of December 7, 1941.

In early 1942, censorship was imposed on domestic dissemination of weather information. Instead of detailed forecasts, the Bureau issued general statements, such as "tomorrow will be warmer," or "today will be a good day to cut hay." Weather maps were not published until at least a week old. Even private citizens were held accountable. When Eleanor Roosevelt inadvertently mentioned rain in California in her newspaper column, she was immediately reprimanded by the Office of Censorship. By 1943 the turning tide of war led to the relaxation of many restrictions.

During World War II, the Department of Commerce's Weather Bureau provided over 700 men to the Armed Services. In addition, Bureau meteorologists served as instructors at civilian colleges and universities and special service schools where military meteorological candidates were given crash courses in the science of weather. Once the program got into full swing, weather observers were trained at the rate of 150-200

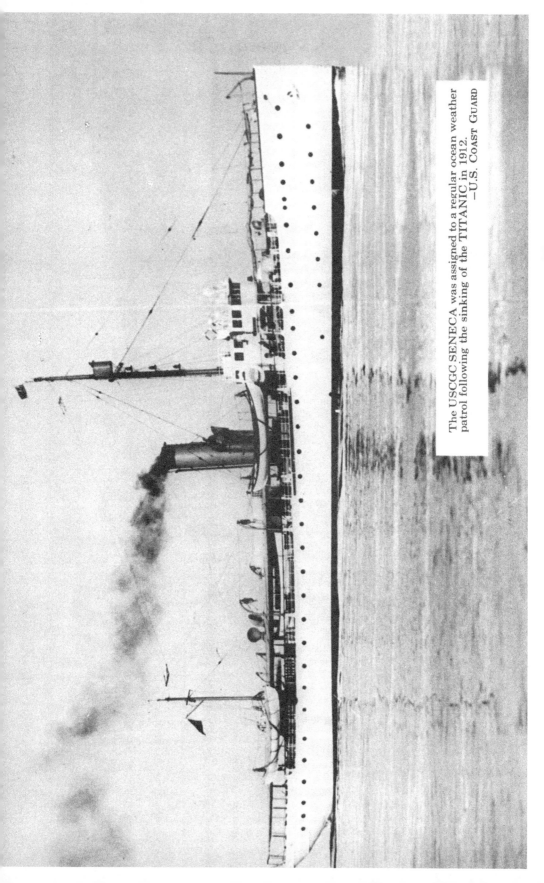

The USCGC SENECA was assigned to a regular ocean weather patrol following the sinking of the TITANIC in 1912.
—U.S. Coast Guard

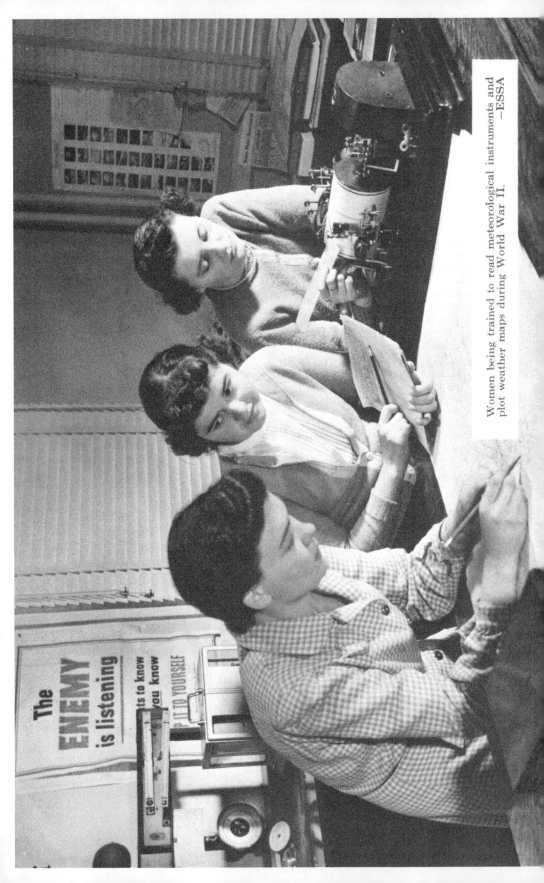

Women being trained to read meteorological instruments and plot weather maps during World War II.

—ESSA

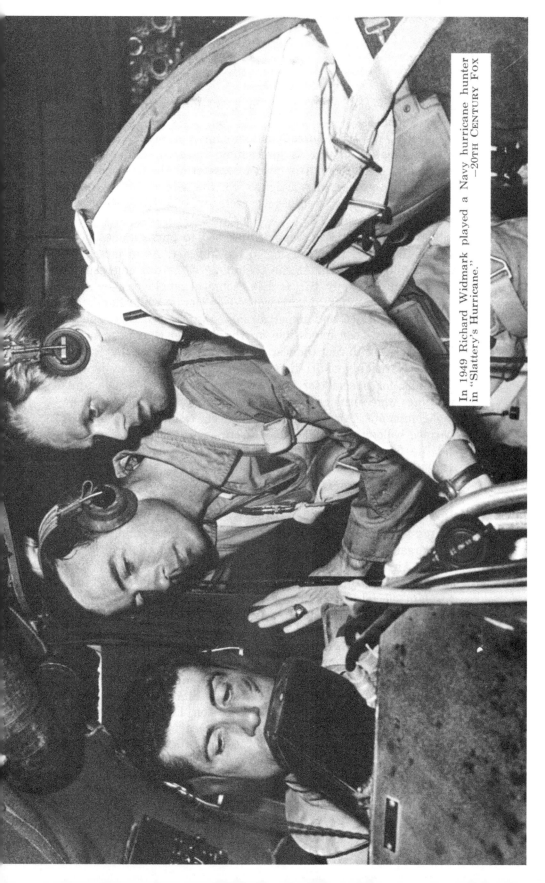

In 1949 Richard Widmark played a Navy hurricane hunter in "Slattery's Hurricane."
—20TH CENTURY FOX

every two weeks, while 2,000 to 3,000 forecasters graduated every 3 months.

Selected Weather Bureau meteorologists also served in the various theaters of war as forecasters, research scientists, advisors, and instructors. The head forecaster of the Chicago District, for example, was assigned to the China-Burma-India Theater and prepared weather forecasts for B-29 strikes throughout Southeast Asia.

The Weather Bureau was hard-pressed to find qualified personnel to replace those serving in the military. As in industry, women filled the need. At the beginning of the war emergency, the Bureau had only two female technical employees; by V-J Day, it had over 900. Many worked as clerks, while hundreds more took surface and upper air observations. Without them, it would have been impossible for the Weather Bureau to function effectively during the war.

The Nation's mobilization for war brought unprecedented demands for special services. Weather Bureau forecasts and warnings were furnished for munitions plants, flying fields, training commands, Army Engineer projects, glider detachments, barrage-balloon and smoke detachments, port authorities, chemical warfare units, aircraft interceptor commands, proving grounds, coast artillery units, military weather installations, and many others.

A large number of munitions plants were built in 1940-41. Because of the huge amounts of high explosives handled, they required protection from lightning. A severe storm warning service was organized by the Weather Bureau in cooperation with the Army Air Forces Weather Service and the Office of Civilian Defense. When thunderstorms threatened, advisories were issued to plant officials so war workers could be evacuated. Warnings were also sent to military installations. By June 1945 nearly 200 storm-reporting networks had been established.

In 1943, the Bureau's hurricane center at Jacksonville, Florida was transferred to Miami and redesignated the Weather Bureau, Air Force, and Navy Joint Hurricane Warning Central. In addition, a new hurricane center was established in Boston. These centers, plus the one in Washington, D.C., provided hurricane warning protection for the entire Atlantic Seaboard. In that same year, Colonel Joseph P. Duckworth and navigator

Lieutenant Ralph O'Hair of the Army Air Forces flew a light, single-engine plane through the maelstrom of a Galveston-bound hurricane into its eerie "eye," that canyon of clouds and stillness at the heart of every mature hurricane. This flight began a new era in hurricane and typhoon tracking and forecasting, and led to the institution of regular reconnaissance flights now flown by both the Air Force and Navy during the hurricane and typhoon seasons.

Japanese balloon-borne incendiary and antipersonnel bombs began to reach the West Coast in 1944, drifting eastward on wind currents at 30,000 feet. The San Francisco Weather Bureau office prepared daily forecasts of these winds over the Pacific and Western States for the Forest Service, worried about fires in great national forests, and for the Western Defense Command, charged with intercepting and destroying the bombs.

The number of both military and civilian flights increased steadily during the war. In June 1943, the Weather Bureau established an experimental Flight Advisory Weather Service (FAWS) unit at New York City's LaGuardia Field to provide special weather service to the CAA's air traffic controllers, and through them to pilots in flight. A second unit was established at Washington, D.C.'s National Airport in September of that year. At the request of the Army Air Force, the program was expanded, and 22 additional units were established during late 1943 and early 1944. By August 1945, more than 26,000 clearances and flight plans for more than 650,000 aircraft were being handled monthly. The Weather Bureau's FAWS units provided the basic meteorological information to the CAA and Army personnel for dissemination to civilian and military pilots.

Overseas, the Bureau supplied personnel, equipment, and climatological data to the Army Quartermaster Corps, which was engaged in an enormous program of testing and developing equipment for use in the many and varied climates in which the Army was fighting. As plans began to be formulated for the invasions of Africa and then Europe, the Weather Bureau collaborated with various oceanographic institutes in the United States to make available to the Army and Navy the information needed regarding sea and swell conditions at the attack points.

In 1940, the Army Air Force and the Navy established weather centers in Washington. When the Weather Bureau's

Analysis Center began operations early in 1942, the military centers were moved to adjacent quarters. This allowed the three facilities to pool their resources, simplified coordination of activities, and helped eliminate duplication of products and services. After the war the Bureau's Center, by mutual agreement, was expanded to serve the needs of all.

Calculated Weather Risks

World War II gave birth to a new kind of climatology. Military planning had to be done on a long range basis and the "calculated weather risks" or climatological odds were essential planning ingredients.

Before the war, climatology was usually concerned with generalized presentations of weather averages, extremes, or totals for geographic locations or divisions; there were few attempts to apply it to the solution of specific problems. Now suddenly, there was a demand for exhaustive weather "probabilities" for operations and projects as complex as an amphibious invasion and as single-purposed as the design of a weapon to withstand jungle humidity. Often the problem involved geographic areas for which little or no climatic information was available; always an answer—the best possible answer—was expected.

To help provide the basic climatic information military meteorologists and planners needed to calculate these weather odds, the Joint Meteorological Committee asked the Weather Bureau to prepare special climatological summaries for use in existing and anticipated theaters of war.

In October 1941, a joint Army-Navy-Weather Bureau project was undertaken to add additional weather data to the daily Northern Hemisphere weather maps for past years. The purpose was to obtain an historic series of maps which could be analyzed in greater detail and used to extend the range of forecasts, both in time and area. By early 1944, a series of daily maps dating back to 1899 was available to Allied military meteorologists. Much of the basic weather information used in this project was placed on punched cards and data listings were made for each map by machine methods.

As early as the 1920's it was realized that the only practical way to handle the vast masses of climatological data being accumulated by the world's meteorological services was to use mechanical methods. In 1934 the President's Science Advisory Board recommended an IBM cardpunching unit for the Weather Bureau, and the Civil Works Administration subsequently provided the money to machine-process, analyze, and summarize many years of domestic and marine climatological data. Later Works Progress Administration (WPA) money made it possible for the Weather Bureau, working in cooperation with the Army, to also compile and analyze surface and upper air observations from about 400 airways weather stations for the period 1928-41. When war came and the WPA was terminated, the Army Air Force took over the financing of this work.

It was early decided that as many as possible of the weather records for strategic war areas should be placed on punched cards, and approximately 60 million cards were prepared for 911 military weather stations (the Weather Bureau had earlier processed some 100 million cards for domestic stations). After the war, the routine punching of all weather observations was inaugurated at Weather Bureau and Army Air Force stations staffed by full-time personnel.

The widespread wartime use of machines to process weather data demonstrated decided advantages over earlier manual methods. The continuation and improvement of these mechanical techniques, and the application of far more sophisticated electronic technology to the practices and products of meteorology characterized the development of the national weather service during the postwar era.

PART IV
TODAY AND TOMORROW

Today and Tomorrow

On July 23, 1969, the day before the splashdown of Apollo 11, a weather satellite and Navy aircraft reconnaissance indicated there would be heavy showers and some thunderstorms in the prime recovery area on the 24th, making the operation more difficult and perhaps dangerous. After coordinating with the Navy's Fleet Weather Central at Pearl Harbor, Weather Bureau space flight support meteorologists recommended the landing area be moved some 200 miles. NASA made the change and Armstrong, Aldrin, Collins, and Apollo 11 came down in good weather.

In meteorology, as in spaceflight, the adaptation of wartime developments to peacetime uses and the continuing postwar technological revolution have completely reshaped man's horizons.

The Dimensions of Change

After the war, radar and the deadly V-2 rocket became weather instruments, one extending meteorology's vision electronically, the other bringing it to the threshold of space. Meteorological rockets led to meteorological satellites, which make it possible to track major storm systems on a global scale. And along with radar, rockets, and satellites came the electronic computer and machine weather forecasting.

Radar, one of the most significant meteorological developments of World War II, allows the meteorologist to "look out the window" over tens of thousands of square miles and to see at night. It is capable of piercing clouds or fog and penetrating one storm to detect others behind it. Its unique 3-D view provides the location, dimensions, intensity, and movement of storms, as well as changes in their character.

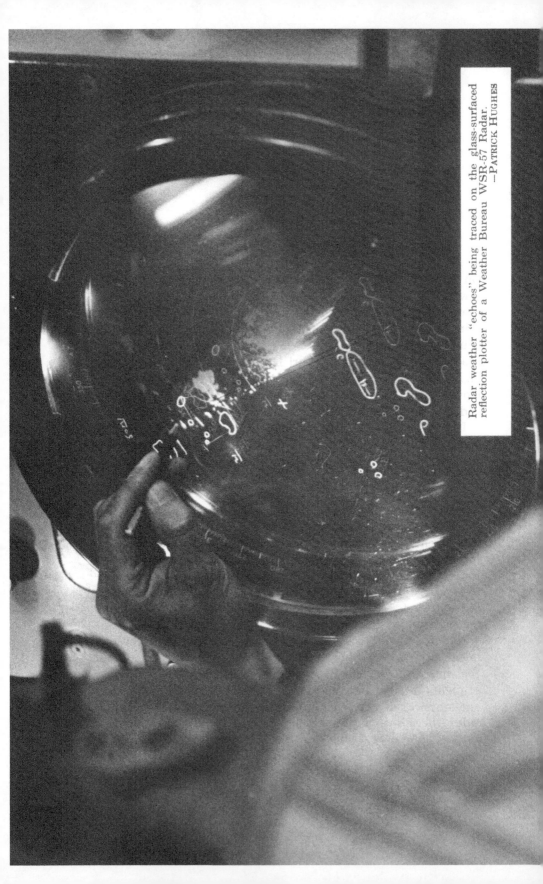

Radar weather "echoes" being traced on the glass-surfaced reflection plotter of a Weather Bureau WSR-57 Radar.
—PATRICK HUGHES

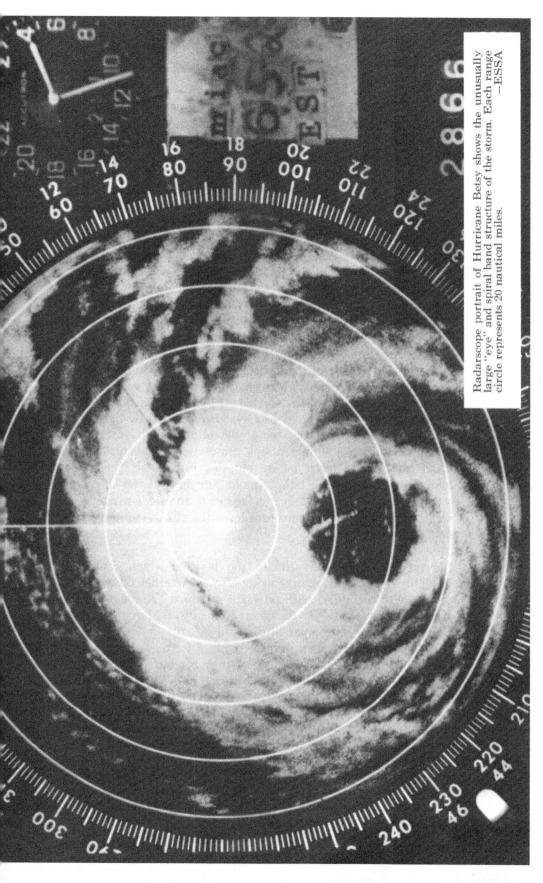

Radarscope portrait of Hurricane Betsy shows the unusually large "eye" and spiral band structure of the storm. Each range circle represents 20 nautical miles. —ESSA

In 1946 the Navy gave the Weather Bureau 25 surplus aircraft radar sets, which Bureau technicians modified for ground meteorological use. Further transfers followed. Meanwhile the Signal Corps developed a radar system specifically designed for weather detection, which was fielded by the Air Weather Services in the early 1950's. In the late 1950's the Weather Bureau developed its own meteorological radar system, which was also adopted by the Naval Weather Service.

Today, the national weather surveillance radar network consists of long range Weather Bureau radars supplemented by Air Force and Navy installations. Most of these are concentrated east of the Rockies to guard the hurricane-vulnerable Atlantic and Gulf coasts, and the tornado-prone midsection of the United States. In the West, FAA air traffic control radars frequently fill in the network, as do air defense radars in Alaska.

There are some Weather Bureau radar installations in the Western States. One of the most unusual is on top of Point Six Mountain, near Missoula, Montana, guarding the vast timberlands of western Montana, northern Idaho, and eastern Washington against the 7 out of 10 forest fires caused by lightning. The radar has proven capable of detecting thunderstorms occurring within a 160-mile radius. Since lightning-struck trees often smolder for hours before bursting into flame, the radar meteorologists can direct the "smoke-jumpers" of the U.S. Department of Agriculture's Forest Service to the areas where such strikes are likely before any fires have the chance to really get going.

"It is difficult to overestimate the benefits which could be brought to mankind by organizing a world weather observation service with the aid of artificial earth satellites." Premier Nikita Khrushchev wrote this to President Kennedy on March 20, 1962. That an earth-orbiting satellite would be an excellent weather observation platform had been recognized for more than a decade before the technology to develop and launch one was perfected. Long before the first weather "bird" flew, American meteorologists were using rockets to probe the upper atmosphere.

When World War II ended, about 100 German V-2 rockets were brought to the Army Ordnance Proving Grounds at White Sands, New Mexico. Here their warheads were replaced with instruments and cameras designed to be parachuted back to earth. The striking, panoramic cloud pictures these rocket soundings produced were the inspiration for the meteorological satellite.

Explorer I, the first United States satellite, was blasted into orbit by an Army Redstone Rocket on January 31, 1958. The first experimental weather satellite was launched on April 1, 1960, and nine more were flown through July 1965 to perfect the system.

The research and development program which produced today's operational meteorological satellite system centered on two spacecraft, TIROS (for *T*elevision *I*nfra*R*ed *O*bservation *S*atellite) and Nimbus, a Latin word for rainstorm or cloud.

The TIROS series was the prototype of today's operational weather satellites. Once the National Aeronautics and Space Administration completed research and development, the Weather Bureau took over funding and management of the TIROS operational satellite system. During the experimental phase, NASA attempted to keep one operating satellite in orbit each hurricane season, for TIROS photographs had proven particularly useful in locating and tracking hurricanes.

Nimbus satellites, larger and more sophisticated than TIROS, are NASA's advanced test platforms for new cameras and sensors. Nimbus I was launched August 28, 1964, and the Nimbus program will continue into the 1970's.

TIROS I was a 270-pound "pillbox" built by RCA under the technical direction of the U.S. Army Research and Development Laboratories (Fort Monmouth, N.J.) and later NASA. During the 78 days its television equipment operated, the satellite took nearly 23,000 pictures of the earth and its shifting cloud cover; about 60 percent of these were meteorologically usable. All 10 TIROS satellites sent back usable pictures. The most striking feature of these photographs was the great degree of organization of large-scale cloud patterns, an organization never apparent from terrestrial and aircraft observations.

Before the advent of meteorological satellites, weather observations were available for less than one-fifth of the globe. Little if any information was received from the polar regions or vast stretches of Asia, Africa, and South America. Over the oceans, tropical storms formed and often grew to maturity undetected

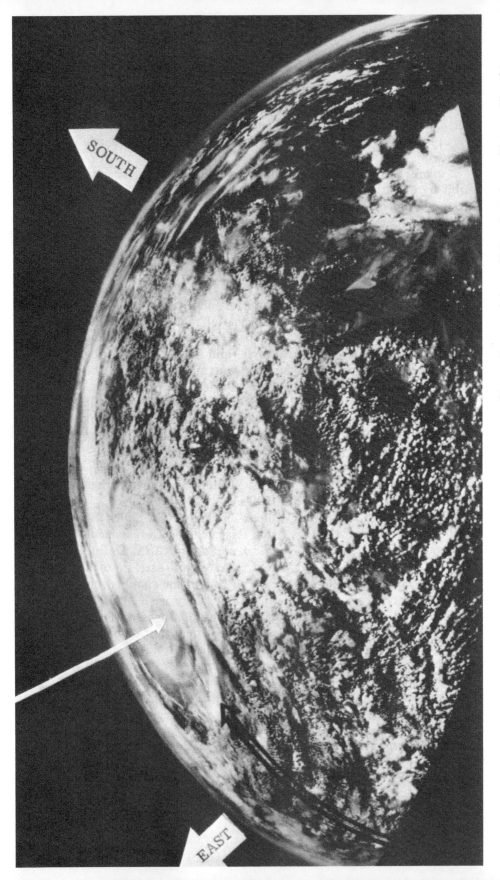

Storm discovered by a V-2 rocket photo. Pictures like this
demonstrated the potential value of meteorological satellites.
—ESSA

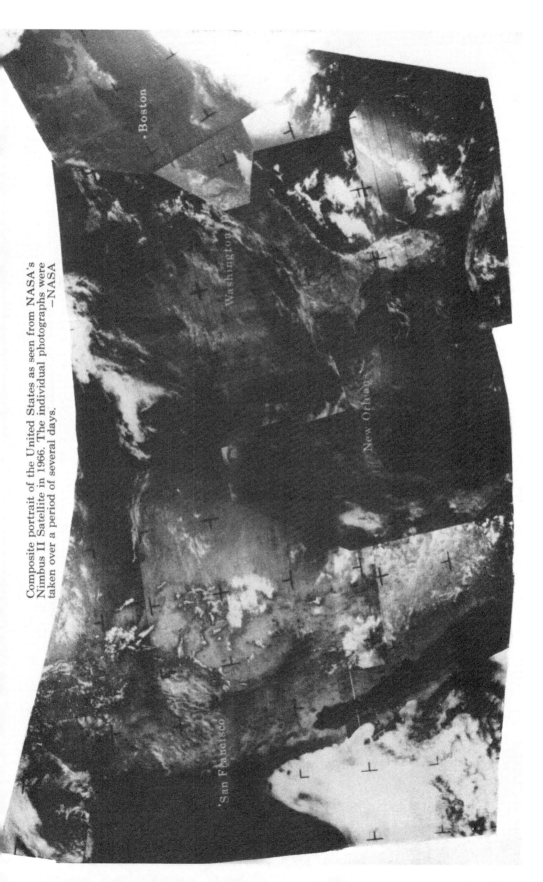

Composite portrait of the United States as seen from NASA's Nimbus II Satellite in 1966. The individual photographs were taken over a period of several days. —NASA

FIRST COMPLETE VIEW OF THE WORLD'S WEATHER

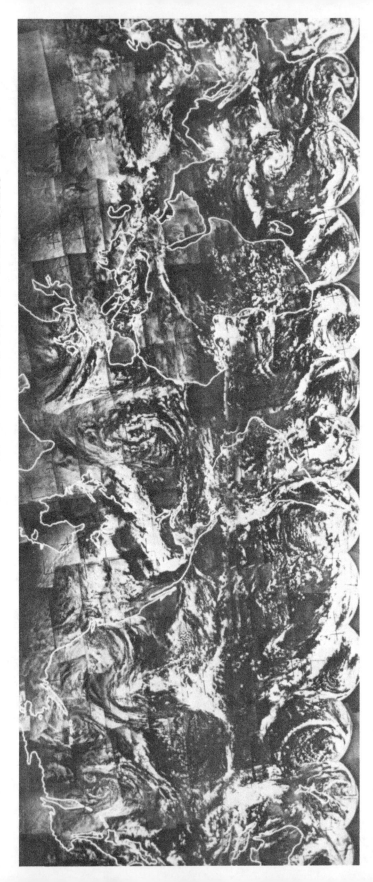

TIROS IX

FEBRUARY 13, 1965

This global photomosaic was prepared from 450 individual pictures. The horizontal white line represents the equator.
—NASA

Part of the air pollution problem facing the United States today.
—ESSA

by man—until they sank a ship or struck an inhabited island or coast.

Although TIROS spacecraft were meant to be experimental vehicles, they provided an unexpectedly large amount of immediately useful data. TIROS I photographed ice conditions in the Gulf of St. Lawrence, and later, on April 10, 1960, detected a tropical cyclone in the South Pacific north of New Zealand. TIROS III tracked Hurricane Anna for several days in July 1961 and later detected the formative stages of Hurricane Esther. A TIROS satellite's warning of the approach of Hurricane Carla in September 1961 made possible the largest mass evacuation in the United States to that time. More than 350,000 people abandoned their homes along the Gulf Coast to flee before the storm.

In August 1960, NASA and the Weather Bureau jointly invited scientists from 21 nations to participate in the analysis of weather data to be gathered by TIROS II. Routine national and international distribution of cloud analyses and storm advisories prepared from satellite photographs was arranged following the launch on November 23, 1960.

In the fall of 1961, NASA and the Bureau began training meteorologists from 27 countries in the use of satellite photographs in weather analysis and forecasting. Later, with the launch of TIROS VIII on December 31, 1963, an automatic picture transmission system was orbited, making it possible for both professional and amateur meteorologists anywhere in the world to receive their own cloud pictures as the satellite passed overhead.

TIROS IX was the first of the experimental series to be placed in polar orbit and to turn on its axis like a wheel, rather than having the satellite's bottom pointing earthward. This orbit and flight attitude are used with today's operational satellites because the combination makes possible the photographic observation of the entire earth during daylight hours. From these pictures a mosaic of the earth's total cloud cover—a true world weather map—can be routinely produced every day.

Lewis Fry Richardson, a British meteorologist, was apparently the first to make a mathematical weather forecast. Working for 10 years, he solved numerous complex equations to approximate atmospheric behavior and finally arrived at a forecast—25 years before the electronic computer made his method practical.

In the mid-1940's the great mathematician John von Neumann and his colleague Dr. Jule Charney of the Institute for Advanced Study at Princeton, N.J., began using early computers to make weather predictions. Much had been learned about the atmosphere since Richardson's pioneer work, and this new knowledge, particularly the simplified equations developed by Carl Gustaf Rossby, made it possible to improve Richardson's technique.

The Weather Bureau, Air Weather Service, and Naval Weather Service, working closely with the Institute for Advanced Study, the Massachusetts Institute of Technology, and the University of Chicago, formed the Joint Numerical Weather Prediction Unit at the Weather Bureau Analysis Center in 1954. By April 1955 operational computer forecasts were being made routinely, and meteorology had entered a new era.

Given a comprehensive picture of the initial state of the atmosphere and a precise mathematical description of its processes, a large computer should be able to make accurate weather forecasts. Unfortunately, the picture is incomplete, our understanding of atmospheric processes rudimentary. Because of this, computer forecasts are far from perfect and have to be limited to a few days; the longer the forecast period, the less the real atmosphere resembles the initial model given the computer.

The complexity of the atmosphere is also a problem. In no other application have the capacities of electronic computers been taxed so heavily as in weather forecasting. Even the simplest approximations of the atmosphere exceeded the capacities of early machines, and today the largest computer commercially available must still handle weather prediction problems in segments.

In developing its own computer forecasting capability, the Navy's Fleet Numerical Weather Central in Monterey, California has adopted an analogue, rather than atmospheric modeling technique. Navy meteorologists use the computer's speed and phenomenal memory to compare today's weather pattern with those of the past. When a match or near match is found, its subsequent development is used as a probable prediction for the current pattern.

The Air We Breathe

In October 1948 industrial waste poisoned the air over Donora, Pennsylvania, killing 20 people. Almost 6,000 more suffered coughing fits, eye inflammation, nausea, and diarrhea. In London in December of 1952 and 1962, severe air pollution was blamed for the deaths of 4,000 and 600 people respectively, and there are some indications that many Londoners today may be more prone to respiratory and heart diseases.

When the air aloft is warmer than that near the ground, the heavier surface air cannot rise to mix contamination through higher layers. When this "lid" is near the ground and surface winds are very light, the drab pall of smoke and smog so characteristic of many of our cities and industrial complexes develops. And there is evidence that this noxious mixture is gradually spreading into a nebulous world-encircling veil of pollution.

In 1955, Public Law 159 authorized Federal assistance to State and local agencies to combat air pollution. Primary responsibility for the program was assigned to the Department of Health, Education, and Welfare's National Center for Air Pollution Control in Cincinnati, Ohio, where meteorological activities were to be conducted by Weather Bureau personnel.

During the late 1950's, Center meteorologists began making experimental forecasts of air pollution potential over the Eastern United States. The success of these forecasts led to the establishment of a routine advisory service in the fall of 1960, a service extended to all the 48 contiguous States 3 years later. On July 7, 1967, responsibility for issuing these advisories was transferred to the Weather Bureau's National Meteorological Center.

In 1968 the Cincinnati Center became the National Air Pollution Control Administration, which in the fall of 1969 moved its offices to Raleigh, N. C. Earlier, in the summer of 1969, the Administration transferred funds to the Weather Bureau to inaugurate special observation programs in St. Louis, Chicago, New York, Philadelphia, and Washington, D. C., to record temperature, pressure, humidity, and wind speed and direction from the earth's surface to about 10,000 feet. This information is interpreted by specially trained air pollution meteorologists, then transmitted to local air pollution control agencies. The Weather Bureau has requested funds to continue these observation programs and to establish similar ones in the Nation's 15 most critical pollution areas.

In 1966 weather officials from 25 nations including the United States met in London for the first International Clean Air Congress. The second Congress will be held in Washington, D. C. in 1970.

Another agency deeply involved in air pollution studies is the U.S. Atomic Energy Commission. Nuclear energy, a partial answer to the problem of combustion-caused pollution, produces a potentially dangerous pollutant of its own: radioactive particles and gases. If released into the atmosphere, this material is carried by the wind, and much of the work of predicting its movement and dilution is based on upper wind patterns and forecasts.

One of the earliest applications of meteorology to such problems was the prediction of the movement and concentration of fallout from atmospheric nuclear weapons tests during the wartime Manhattan Project. Forecasting techniques were considerably refined during subsequent tests at Bikini Atoll in 1946 and Eniwetok in 1948, where Air Force meteorologist Brig. General Benjamin Holzman (retired), then a Colonel, was responsible for the fallout predictions.

In 1947, control of nuclear energy was vested in the newly created civilian Atomic Energy Commission. In addition to its own meteorologists, a staff of Weather Bureau meteorologists has been working closely with the Commission since 1947. In 1955, the Weather Bureau accepted responsibility for operating the meteorological observatory and research unit at the Nevada Test Site near Las Vegas, where initial meteorological support had been provided by the Air Weather Service.

The United States has not tested nuclear devices in the atmosphere since 1962, and fallout prediction is no longer the major focus of the AEC program. However, because of the possibility of accidental leakage of radioactivity, Weather Bureau meteorologists still supply forecasts, particularly wind and rain predictions, in advance of all underground tests. Weather forecasts are also an important factor in conducting the AEC's "Plowshare" experiments on the use of nuclear explosives as peacetime engineering tools. Weather Bureau meteorologists

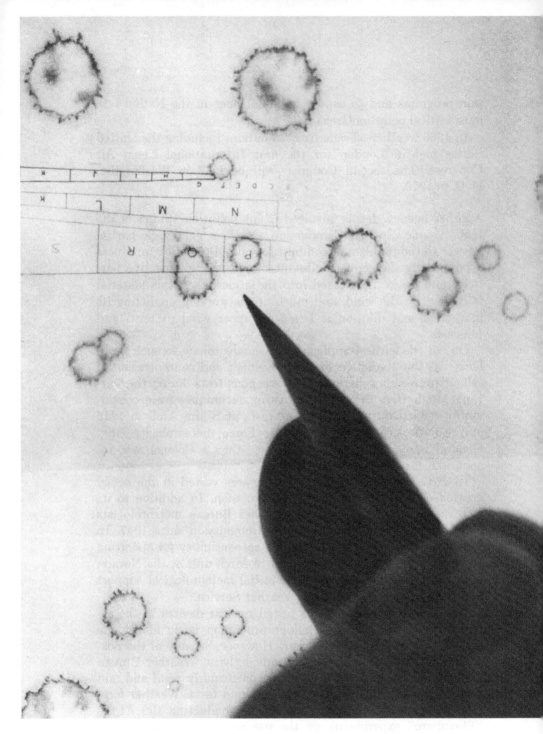

An important program sponsored by the Atomic Energy Commission involves evaluating the efficiency of rain and snow in removing particulates from the air. Here raindrop samples are collected on sensitized paper to determine their size. (The spots are actually several times larger than the raindrops that made them, but a correction factor has been established in laboratory tests.) Size distributions of raindrops obtained by this and other methods are used to predict "washout."

–ATOMIC ENERGY COMMISSION

conduct research and provide support services for the AEC in Washington, D. C., as well as at its nuclear testing station near Idaho Falls, Idaho and the Oak Ridge National Laboratory in Tennessee.

A significant cooperative research effort between the AEC and the Weather Bureau preceded the commissioning of the Shippingport, Pennsylvania nuclear reactor in 1957, the first reactor to generate commercial electric power. Meteorology, particularly knowledge of climatological patterns of wind and weather, is an important consideration in the location of nuclear power plants.

Through the years, the Atomic Energy Commission has sponsored a broad range of meteorological research at its various national laboratories, and by numerous universities and private research organizations. Many of the techniques used today in conventional air pollution programs were originally developed for research on radioactivity. Recently AEC laboratories have turned more of their attention directly to the problems of non-radioactive air pollution in cooperative programs with other government agencies such as the Department of Health, Education and Welfare. Experience and techniques gained in AEC research are being applied to such problems as "fingerprinting" and identifying the sources of industrial air pollution and developing automatic pollution monitoring and warning systems for heavily populated areas.

Age of the Interface
Today, the former Weather Bureau meteorologists assigned to the Atomic Energy Commission and the National Air Pollution Control Administration work for the Air Resources Laboratory of the Environmental Science Services Administration (ESSA). The atmosphere, the oceans, and the earth itself are interacting, not independent elements of man's natural environment; to understand and predict the behavior of one, he must study all three. As man's technology and its applications grow ever more complex, so does the need for coordination of the sciences concerned with his physical environment.

In May 1965, President Johnson submitted a Commerce Department reorganization plan to Congress. It established a new agency to ". . . provide a single national focus to describe,

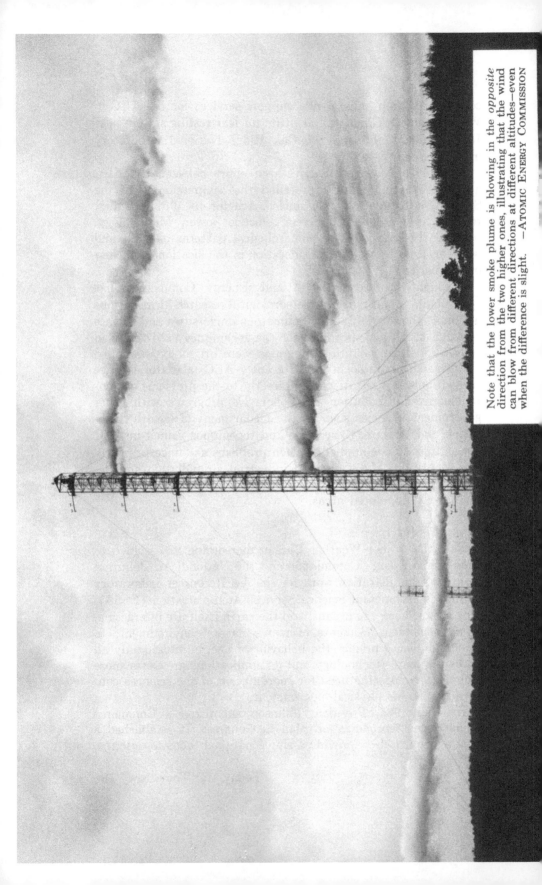

Note that the lower smoke plume is blowing in the *opposite* direction from the two higher ones, illustrating that the wind can blow from different directions at different altitudes—even when the difference is slight. —ATOMIC ENERGY COMMISSION

understand, and predict the state of the oceans, the state of the upper and lower atmosphere, and the size and shape of the earth." The Department of Commerce's Environmental Science Services Administration was officially chartered on July 13, 1965.

The new agency merged the organizations and duties of the Weather Bureau, Coast and Geodetic Survey, and the Central Radio Propagation Laboratory of the Bureau of Standards. Both the Weather Bureau and Coast and Geodetic Survey remained as organizational entities under ESSA. In addition, three sister components were created: the National Environmental Satellite Center, the Environmental Data Service, and the collective Institutes for Environmental Research, now ESSA's Research Laboratories.

Dr. Robert M. White, an authority on atmospheric dynamics who succeeded Dr. Reichelderfer as Chief of the Weather Bureau in 1963, was appointed Administrator of ESSA. Dr. George P. Cressman, a pioneer in numerical weather prediction, became Director of the Weather Bureau.

Under ESSA, the Weather Bureau still provides the Nation's basic weather services and warnings. Its National Meteorological Center is the composite successor of the Bureau's wartime Analysis Center, its Extended Forecast Section, and the Weather Bureau, Air Force, and Navy Joint Numerical Weather Prediction Unit. The National Center provides the analyses and guidance forecast material used by Bureau meteorologists in the field to prepare local forecasts. Its products include forecasts for up to 48 hours made four times a day, and 72-hour forecasts made once a day. Five-day extended forecasts are currently issued three times a week, and 30-day outlooks twice a month.

Since 1958, incoming meteorological data have been fed directly from the teletype circuits into a Center computer, eliminating manual processing. The introduction of this procedure and of improved computer models of the atmosphere, as well as a shift from North American to Northern Hemisphere analysis and forecasting and the continuing increase in computer speed (at a rate of 50% a year), have significantly improved the quality of the Center's short range and 5-day guidance forecasts. This material is now available to the local forecaster several hours earlier than was before possible.

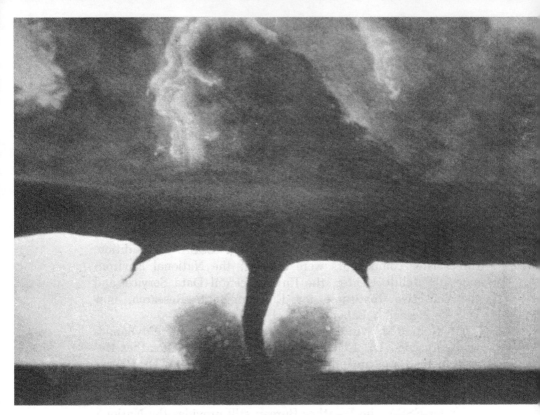

The first known photograph of a tornado, taken 22 miles southwest of Howard, South Dakota on August 28, 1884.
　　　　　　　　　　　　　　　　　　　　　　　　　　−ESSA

A milestone was reached in 1966 with the introduction of a new, six layer mathematical model of the atmosphere capable of producing longer range forecasts. The Center expects to begin issuing computer-generated 5-day forecasts during 1970.

Major advances have been made in severe storm warning services since World War II. Although a Signal Corps Lieutenant, J. P. Finley, prepared practice severe storm alerts as early as 1884, the ability to issue reasonably detailed predictions and warnings had to await the evolution of modern upper air observation methods and the development of radar. Until the early 1950's, statements such as, "conditions are favorable for severe local storms this afternoon," was about as far as the Weather Bureau would go in the way of an official alert. The word "tornado" was never used for fear of public panic.

In 1948 two Air Weather Service meteorologists, Major Ernest J. Fawbush and Captain Robert C. Miller, developed practical techniques to forecast the likelihood of tornado development.

Tornadoes are most deadly when unexpected. They can and do strike in all 50 States. —ESSA

The Weather Bureau also issues warnings for winter storms such as this one which stranded several trains in drifted snow in 1966.   —NORTH DAKOTA STATE HIGHWAY DEPARTMENT

They issued their first warning for Tinker Air Force Base, Oklahoma on March 25 of that year; a tornado struck the base at 6:10 that evening.

The success of subsequent warnings led to creation of an Air Force severe weather warning center at Tinker Air Force Base in 1951. The following year the Weather Bureau established a severe local storms forecasting unit in Washington, D. C., which moved to Kansas City, Missouri in 1954 and was joined by the Air Force group in 1956. After a few years of joint operation, severe weather forecasting responsibility was assumed by the Weather Bureau in 1961.

Today, the Bureau's severe storm service provides warnings of tornadoes, severe thunderstorms, hail, and damaging winds. Air Force and Weather Bureau meteorologists still work together, the Air Weather Service alerting military personnel and the Weather Bureau, the general public.

Bureau meteorologists make no claim that they can predict exactly when and where a tornado will strike. If atmospheric conditions appear to favor severe thunderstorms or tornado development, they issue a watch outlining the area where development is likely. When such a watch is issued, local weathermen alert their networks of volunteer storm spotters. If the threat is confirmed visually or by radar, the local Weather Bureau office issues a community warning describing the location, movement, and probable path of the storm or storms.

On Palm Sunday 1965, the Weather Bureau gave warnings up to several hours in advance for 33 of 37 tornadoes that ripped through much of the Midwest. Unfortunately, these warnings were often ignored and 266 people were killed and another 3,261 injured. Said one man whose house was damaged, "We were watching the Walt Disney show when they interrupted the program for a tornado warning. We should have realized it was important if they interrupted Walt Disney, but we'd never been in a tornado and just didn't pay any attention."

As a result of the Palm Sunday tornadoes, a national Natural Disaster Warning (NADWARN) system was established, with ESSA primarily responsible for its implementation. Ultimately, NADWARN will provide a round-the-clock rapid warning service to every American community with a radio or television station.

Ice storms, with their transparent beauty but deadly glaze are another winter weather hazard of concern to Bureau forecasters.
—ESSA

Captains Marvin A. Lillie (left) and Robert Y. Foerster, aircraft commanders of the 53rd Weather Reconnaissance Squadron's Hurricane Hunters, received the Distinguished Flying Cross for their perilous flights into Hurricane Camille on August 17, 1969. (Their 13 aircrewmen all received the Air Medal.) Captain Lillie's plane lost two of its four engines on its second flight, while Captain Foerster and his crew logged a total of 11 hours in Camille, whose surface winds were at times estimated at over 200 m.p.h.   –U.S AIR FORCE

The National Hurricane Center in Miami is the nerve center of the hurricane warning service. Local warning responsibility is shared by the Miami center and Weather Bureau offices in New Orleans, Washington, Boston, and San Juan. Offices in San Francisco and Honolulu issue similar alerts and warnings for tropical cyclones in the eastern Pacific.

Since World War II, aircraft reconnaissance has become a vital tool in hurricane surveillance. Air Weather Service "Hurricane Hunters" operate out of Ramey Air Force Base, Puerto Rico, while their "Typhoon Chasers" (hurricanes are called typhoons in the western Pacific) are based at Guam. The Navy's counterparts are Airborne Early Warning Squadron One and Weather Reconnaissance Squadron Four. VW-1 is based at

Guam, VW-4 at Jacksonville, Florida, with an advanced detachment stationed at Roosevelt Roads, Puerto Rico, during the hurricane season.

In the spring of 1969, massive Midwest floods caused $100 million damage and forced 25,000 people from their homes. Yet, only eight people were killed. Precise flood forecasts made long in advance saved countless lives and an estimated $250 million.

Because the behavior of a river is limited by the character of its basin, river forecasting is considerably more accurate than weather forecasting. The tools used by Weather Bureau hydrologists to make river forecasts are rain, snowfall, and evaporation measurements, river gage readings, and more recently, computer models of the river basins themselves.

Principal river basins in the United States are served by a Weather Bureau river forecast center supported by river district offices and an extensive network of substation and volunteer observers. Forecasts which predict flood crests well in advance enable local officials to plan levees and, if necessary, evacuations. These forecasts are distributed to cooperating agencies such as the Army Corps of Engineers—the chief flood control agency of the Government—the Red Cross, Civil Defense, the Office of Emergency Planning, local levee boards, and many others. The backbone of the service are the 4,000 volunteer river and rainfall observers who provide the basic observations essential to the forecaster.

Flash floods are different. Occurring suddenly after an unusually heavy rain or a sudden snow melt, they are difficult to predict and can occur almost anywhere there is a river, stream, or even dry gulley. The meteorologist's ability to predict a flash flood is extremely limited, and the Weather Bureau can only warn of their possibility over large areas. On streams with a high potential for flash flooding, volunteer warning networks are run by local forecasters. Like the observers in similar tornado networks, these volunteers give their neighbors the precious gift of a few minutes warning.

Nationwide, Weather Bureau agricultural services continue much as they were prior to World War II. The Bureau, however, has become increasingly aware of agriculture's need for more specialized programs, and a pilot project undertaken in the

The Coast Guard's Ambrose offshore light structure at the entrance to New York's Harbor provides meteorological observations to the Weather Bureau. The tower replaces the Ambrose Lightship (background), shown here in 1967 about to make her final departure.   —U.S. COAST GUARD

Mississippi Delta in 1959 has become the prototype for tailored services now provided in 12 weather-sensitive agricultural areas. As funds become available, this program will be extended to the rest of the country.

The fire weather service encounters its major season in most areas between July 31 and September 30. Since World War II new fire weather districts have been established in Michigan and Tennessee, and the Weather Bureau plans to gradually extend the service to the entire Nation except Hawaii, where forest fires are not much of a problem.

More than 2,000 U.S. flag merchant ships participate in the cooperative marine meteorological observation program, and some 900 broadcast their information to the Weather Bureau for immediate use by the forecaster. All send their observations to ESSA's National Weather Records Center in Asheville, N.C., where they are computer-processed for climatological studies. A corresponding, international effort involves 6,000 ships of all flags which observe and report the weather across the global sea. In addition, weather observations are received from the Navy's major fleet vessels, large Coast Guard cutters, Coast and Geodetic Survey and other oceanographic research ships, and a limited number of buoy-mounted automatic weather stations, lightships, and offshore towers.

Weather Bureau high seas forecast offices in Washington, San Francisco, and Honolulu issue regularly scheduled marine weather and sea state reports and forecasts for roughly the western half of the North Atlantic and most of the North Pacific. Other offices provide forecasts and warnings to the fishing fleets that operate in the western North Atlantic, the Caribbean, and the Gulf of Mexico.

For mariners who stay closer to land, small craft, gale, storm, and hurricane warnings are broadcast for the coastal waters of the United States and its territories, inland waterways, and the Great Lakes. Visual displays—flags by day and lights by night—are operated by the Coast Guard, Weather Bureau, State and local governments, and private interests for mariners lacking radio equipment. On the Great Lakes the service includes annual predictions of the beginning dates for ice-free navigation into selected ports, while in some areas surf, sea state, and sea surface temperature forecasts are becoming routine.

ESSA 3

The ESSA 3 meteorological satellite. In operation this space-craft rolls along on its axis in polar orbit.   —ESSA

Marine weather reports, warnings, and forecasts are broadcast by the Navy and Coast Guard. In addition, the Weather Bureau operates automatic telephone briefing systems at eight coastal locations, and more than 2,000 commercial radio and television stations broadcast marine weather information several times a day.

The average weather conditions over the world's oceans are computed at the National Weather Records Center and presented graphically in the Naval Oceanographic Office's series of *Pilot Charts*, the *Climatological and Oceanographic Atlas for Mariners* prepared by the Oceanographic Office and ESSA's Environmental Data Service, and the Naval Weather Service Command's *U.S. Navy Marine Climatic Atlas of the World*. *Sailing Directions* and *Coast Pilots*, published respectively by the Oceanographic Office and ESSA's Coast and Geodetic Survey, describe the climates of the world's coastal waters.

Prior to 1940, U.S. transatlantic aviation consisted of two weekly flights to Europe, both by Pan American World Airways. When World War II ended international aviation grew exponentially.

Aviation meteorologists first raised their forecast ceiling to 25,000 feet for the DC-7 and Constellation, then to 45,000 feet for jet aircraft. At the jet's flight altitudes, pilots, passengers, and weathermen were introduced to winds over 200 knots and to the now familiar problem of clear air turbulence.

Today, Weather Bureau aviation forecast centers at Anchorage, Honolulu, New York, San Francisco, Miami, and San Juan serve the weather needs of international civil aviation, while Bureau forecast offices and hundreds of FAA and Weather Bureau stations serve domestic needs. Approximately 70% of the commercial carriers flying the Atlantic use computer generated flight plans based in part on the computer wind and temperature forecasts issued by the Bureau's National Meteorological Center. These forecasts are transmitted directly from computer to computer over high speed communication lines.

Weather observations solely in support of aviation are made at over 400 civilian and military stations in the United States, with the FAA providing by far the majority of civilian reports. At many stations, the meteorological program is a coordinated effort between the Bureau and FAA or one of the military weather services.

The supersonic transport of the 1970's will pose new and vexing weather problems for civilian meteorologists who will have to worry about temperature, wind, turbulence, and even cosmic radiation at altitudes up to 100,000 feet, problems long faced by military weathermen. They will also be concerned with the intensity of the surface sonic boom, a critical factor in SST use.

The National Environmental Satellite Center manages the operational meteorological satellite program. The first ESSA (Environmental Survey Satellite) spacecraft was launched on February 3, 1966, carrying two television cameras with picture storage capability. ESSA 2, launched February 28, 1966, carried two automatic picture transmission (APT) cameras. In both types of satellite one camera is used operationally, the other held in reserve. The national operational meteorological satellite system—which calls for both types of satellite in orbit at the same time—formally came into being March 15, 1966, when NASA transferred control of ESSA 2 to the National Environmental Satellite Center.

The Center controls the ESSA spacecraft through an operations center at Suitland, Maryland, and command and data acquisition stations at Gilmore Creek, Alaska, and Wallops Island, Virginia. Beginning with ESSA 3, picture storage satellites have also been equipped with sensors to measure radiant energy reflected and emitted by the earth.

Stored data satellite photographs are computer-processed into global cloud maps and transmitted to weather stations throughout the world. When necessary, special warnings of severe storms are sent to the meteorological services of nations in threatened areas. By a 1962 agreement, the United States and the U.S.S.R. exchange satellite data over a Washington-Moscow weather wire. To date, the Russians have furnished information from their COSMOS 122 satellite, launched in 1966, and COSMOS 144, 156, 184, 206, and 226, launched in 1967 and 1968.

The automatic picture transmission system developed by NASA enables a satellite to transmit cloud cover pictures directly to simple ground and shipboard stations anywhere in the

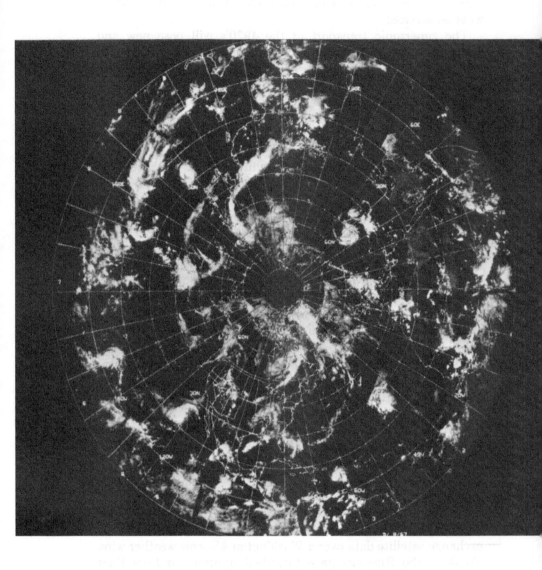

This computer generated mosaic of cloud photographs taken by ESSA 5 shows more than a dozen storms, including five hurricanes.   –ESSA

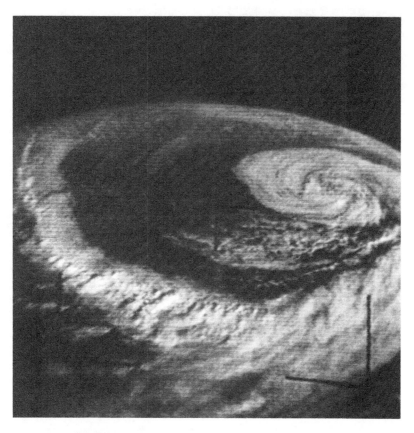

Satellite picture of a storm system in the North Atlantic Ocean southeast of Nova Scotia on May 29, 1963. The long cloud spiral is associated with a weather front at the surface.
—MONTHLY WEATHER REVIEW

world. Receivers can be built for under $500 or procured commercially for as little as $5,000. Today's APT users include the meteorological services of 45 countries, more than a score of universities, as many television stations, and an unknown number of private citizens who have built their own receivers.

By interpretation of satellite data it is possible to determine snow and ice cover, and summaries of snow and ice fields are now prepared weekly. Photographs of the snow cover in the Western States are being used in experimental support of the Weather Bureau's river and flood forecasting service.

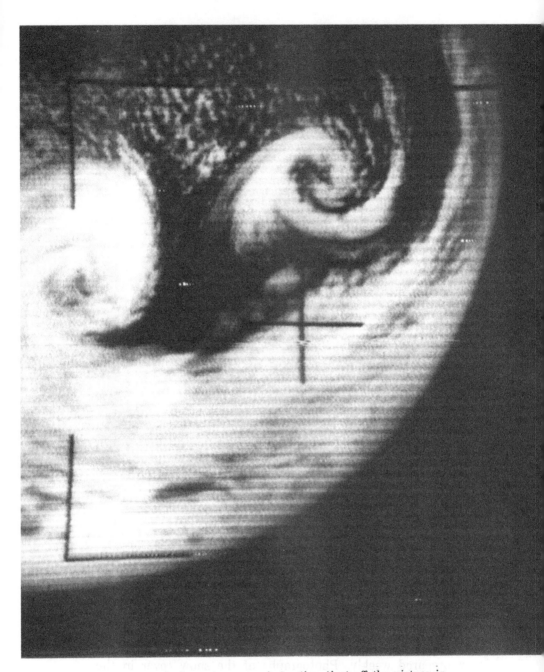

Two cyclonic centers near Antarctica (just off the picture in the lower left). Meteorological satellites provide weather data for the Southern Hemisphere, where few observations were previously available.   —MONTHLY WEATHER REVIEW

An Air Force weather officer in Vietnam pieces together a satellite weather map received on a facsimile machine at Cam Ranh Bay Air Base, Vietnam. —U.S. AIR FORCE

Satellite data provide invaluable meteorological information for the Southern Hemisphere, where few weather observations are otherwise available. This was a factor in the Navy's decision to discontinue stationing weather ships along the air route from Christchurch, New Zealand to Antarctica in support of Operation DEEPFREEZE. In addition, APT readouts support fleet operations, as well as Army and Air Force activities such as worldwide photomapping and aviation forecasting.

As this book went to press, TIROS M, an experimental prototype for an improved ESSA satellite series, was scheduled for launch in November 1969. It will provide nighttime cloud coverage, previously lacking, and will also combine stored data and APT readout capability in a single spacecraft. In 1972, NASA plans to launch another satellite whose orbit will make it appear stationary over the Equator. Such a satellite will be able to follow the life cycle of short-lived storms that might be missed by a globe-circling spacecraft.

Another development of interest to meteorologists is the experimental color camera carried by NASA's application technology satellite 3, launched in November 1967. Much more useful meteorological information can be gleaned from color pictures than black and white because of the color contrast between cloud systems and geographic features. There is even some evidence that such pictures can indicate storms of tornado-generating intensity.

A 91-pound satellite infrared spectrometer called SIRS A was flown on NASA's experimental Nimbus III satellite in April 1969. Built by ESSA's Satellite Experiment Laboratory, the instrument provides a profile of air temperature at various levels between the spacecraft and the ground or the top of a cloud layer.

SIRS A currently makes 5,000 temperature soundings per day in the Southern Hemisphere alone, whereas conventional instruments furnish only about 70 daily upper air soundings from that half of the globe. Analyses of atmospheric circulation patterns based on SIRS A data agree with those based on conventional upper air observations.

Long before SIRS A began orbiting the earth, the Satellite Experiment Laboratory was working on its descendants. SIRS B, scheduled to be flown in 1970, will measure both temperature

Meteorological rockets are used to probe the boundary of
atmosphere and space.   —NASA

and water vapor; these measurements can be combined to esti-
mate an atmospheric humidity profile. Sounders based on the
SIRS series will become part of the standard equipment on
ESSA operational satellites during the early 1970's.

The availability of global data from satellites does not eliminate the meteorologist's need to know the fine structure details
of the atmosphere and of the important linkages between the
upper atmospheric boundary and the denser layers below, knowledge now available only by balloon and rocket-borne radiosondes.

Rockets boost radiosondes to high levels, where the instruments separate and parachute earthward, broadcasting a wind
and weather profile as they descend. Rocketsonde measurements
provide a special type of meteorological information—such as

Today and Tomorrow   155

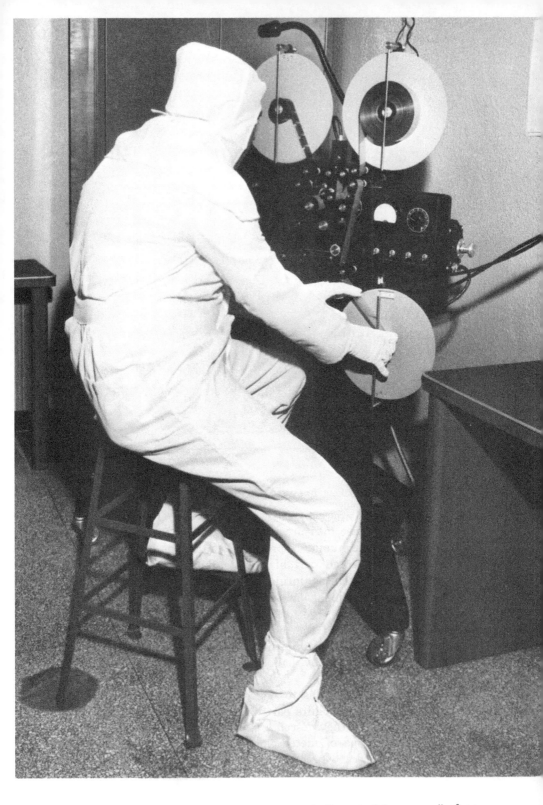

The National Weather Records Center's "clean room," where satellite photographs and microfilm of weather records are processed and duplicated. —ESSA

the density of the atmosphere where a tiny three-man capsule, going 25,000 miles per hour, reenters on its way back from the moon.

The Environmental Data Service archives and disseminates the geophysical observations made by ESSA and collected from other sources, and also manages the Weather Bureau's former climatology program. The National Weather Records Center in Asheville, North Carolina, its largest data center and the successor to the Weather Bureau's climatological data processing unit established in the 1930's, is the principal source of meteorological and climatological information in the United States.

The Center's holdings range from 18th century diaries and journals to the more than 100 million weather observations now received annually from a global observation network. To meet user requirements, much of the data are summarized in routine publications of global, national, and local focus. The Center also provides microfilm, punched cards, magnetic tapes, radar and satellite film, original weather records or duplicate copies, and automatic instrument traces. Special hand and machine summaries and tabulations are available on request. Most of this information is available at nominal cost. Besides the general public, customers include Federal, State, and local agencies, the scientific and engineering community, and some of the Nation's largest corporations.

Under a reimbursable contract, the National Weather Records Center processes the Navy's weather records and serves many of its climatological needs. The Air Force's Environmental Technical Applications Center in Washington, D. C., which provides climatological services to both the Air Force and Army, has its data processing division in Asheville. There is a complete interchange of climatic records between ESSA and the military services.

Climatology, an indispensable tool to the military planner during World War II, had even greater application when peace came. The returning G.I., the farmer, architect, engineer, manufacturer, retailer, and the housewife could all use "calculated weather risks" in their own planning.

Immediately after the war there was a housing boom, and the starting point for building construction at any location is the climate. The Nation's billion dollar airport building program also needed climatological data to choose sites and to determine the number, direction, and length of runways, the need for drainage and for ice and snow removal, the installation of landing equipment, and the location and construction of buildings. In addition, climatic data were required for planning the most economical and safest air routes and schedules, as well as for the design of the aircraft themselves.

In 1954 Dr. Helmut E. Landsberg became Director of the Weather Bureau's Office of Climatology and, when ESSA was formed, the first Director of the Environmental Data Service. Dr. Landsberg guided the development of the current national climatological service. When he retired in December 1966, he was succeeded by Dr. Woodrow C. Jacobs, a former Director of Climatology for the U.S. Air Force.

Today climatological data is used in planning for practically every phase of human activity from recreation to spaceflight. The Environmental Data Service's State Climatologists provide local counsel on the application of climatology to the problems of the community.

Other Weathermen
Although ESSA is responsible for the Nation's basic weather services, many other Federal agencies have related meteorological missions. To prevent duplication of effort and coordinate Government activities, the Department of Commerce established the Office of the Federal Coordinator for Meteorology in 1964. Dr. Robert M. White, then Chief of the Weather Bureau, was appointed to the post. Two committees—the Interdepartmental Committee for Meteorological Services and the Interdepartmental Committee for Applied Meteorological Research—provide policy guidance for the Coordinator.

When the Weather Bureau was transferred to the Department of Commerce in 1940, the Department of Agriculture retained authority to make snow surveys and conduct research related to weather and crops, long range forecasting, and the relationships between weather and soil erosion.

Mountain snow is the great natural reservoir of the West. A large portion of the annual flow of such western rivers as the

Columbia, Missouri, Colorado, and Rio Grande comes in the spring and summer when the snow melts. This usually occurs from April to September, the period of least rainfall and the season of greatest crop need for water. It is also the season which sometimes brings widespread destructive floods.

For almost 50 years forecasts have been made of this water supply by measuring the water content of winter snow accumulations. Western snow surveying and water supply forecasting as a federally coordinated operation was first undertaken in 1935, near the climax of the severe droughts of the 1930's. The present highly cooperative snow surveys are coordinated by USDA's Soil Conservation Service. Cooperators include State engineering offices and agricultural experiment stations, municipalities, irrigation interests, and most of the large power companies, as well as the Weather Bureau, Army Corps of Engineers, Forest Service, National Park Service, Bureau of Sport Fisheries and Wildlife, Bureau of Reclamation, Geological Survey, and the Office of Indian Affairs. The Water Rights Branch of the Department of Lands, British Columbia, sponsors a supporting program on the Columbia River basin in Canada.

Each Western State involved fits the snow survey data into special formulas developed to predict the water supply in the snowmelt season. These forecasts are prepared for key river points and for reservoirs in all important watersheds.

Since most snow measuring sites are in high mountains, surveyors must be trained to avoid avalanches and to survive in blizzards, accidents, or other emergencies. Where a snow course is too hazardous or costly to reach, surveyors record snow depth by observing graduated snow markers from an airplane. In the near future snow measurements will probably be made by sensors connected to electronic telemetry systems for relay to base stations.

The Department of Agriculture's role in agricultural weather service is twofold. It investigates the effects of weather on crop production to define agriculture's needs and, in cooperation with ESSA's Weather Bureau and Environmental Data Service, recommends methods of increasing yield or minimizing weather losses. The Departments of Agriculture and Commerce issue joint weekly crop weather reports for each State, as well as the national *Weekly Weather and Crop Bulletin*. In addition,

Lightning-caused fires annually destroy large forest areas in the United States.
—U.S. Department of Agriculture

weather information is carried in nearly all the USDA's Federal-State market news bulletins.

Project Skyfire is the oldest continuous weather modification program in the United States. Headquartered at the Northern Forest Fire Laboratory of the Department of Agriculture's Forest Service in Missoula, Montana, the Project is designed to determine if cloud seeding can suppress lightning and prevent forest fires.

Lightning is a major cause of forest fires in many parts of the world, particularly in North America. In the United States in years of intense lightning activity and dry weather, 15,000 or more such fires may occur. The problem is greatest in the West, where lightning starts more than half of all fires.

While preliminary results of Project Skyfire have been encouraging, a longer experimental period is needed to determine if man can indeed significantly alter the lightning characteristics of Rocky Mountain thunderstorms.

The Geological Survey and Bureau of Reclamation, two agencies of the Department of Interior, are concerned with the science of hydrology—the study of water in the land phase of the hydrologic cycle. Hydrology is intimately related to meteorology, which involves the atmospheric phase of the hydrologic cycle.

The Geological Survey was created by Congress on March 3, 1879. Major John Wesley Powell, the Survey's second director and a pioneer explorer of the West, proposed making topographic maps of Western river basins, and by measuring streamflow, evaporation, and precipitation at key locations, determining the amount of water that would be derived from each square mile of drainage basin. A camp was established at Embudo, New Mexico on the Rio Grande to train a small group of engineers to do the job.

In the 1880's Powell, passionately interested in reclaiming the "Great American Desert," pressed for national aid for irrigation, and in 1894 was given money to organize an irrigation survey within the Geological Survey. The Reclamation Act of 1902 elevated irrigation investigations from a minor to a major effort, and in 1907 the Reclamation Service was formed as a separate agency of the Department of Interior. It became the Bureau of Reclamation in 1923.

John Wesley Powell talking to a Paiute Indian in Arizona in 1873. From 1870 to 1879 Powell directed the survey and mapping of the "Plateau Province" opened up by his 1869 expedition.
—The Smithsonian Institution

Today, Geological Survey stream, lake, and canal gages are located at 17,000 sites and ground water conditions sampled at 30,000 other locations. Water quality is sampled at 2,500 surface water gaging sites and 2,500 of the ground water wells. Precipitation is also measured at many sites, while snow depth and water content are measured at 240 snow courses.

In the last decade the Survey has installed recording rain gages at many small basin stream gaging stations. The information gathered is used to develop a computer model of the rainfall runoff process in each basin, which can be used in conjunction with long term precipitation records from nearby weather stations to synthesize approximate long term runoff records.

Glacier ice covers at least 200 square miles of the Western United States and about 20,000 square miles of Alaska. The volume of water stored in glaciers is many times that in all the lakes, ponds, rivers, and reservoirs of North America.

Meltwater from glaciers (and snow) behaves in a contrary but useful manner. During hot, dry years when the demand for water is greatest, increased water flows from glacial ice storage; during cool, wet years storage is increased and streamflow smaller. The U.S. Geological Survey has been studying glaciers for many years to establish the amount of water storage and to extend man's knowledge of glacier mechanisms and of past climates and water supplies.

Dates of recent glacier fluctuations—as well as of floods and droughts—can be deduced from tree rings, which are useful in constructing a hydrologic history where there are no records and in estimating trends and variability of rainfall, streamflow, and climatic change. A 1964 Geological Survey study of trees along the Potomac River, for example, makes it possible to date local flood occurrences over the past 100 years.

Because snow and ice are such important sources of water in the West, the Survey conducts cooperative research programs with other Government agencies and with several western universities to increase the streamflow from these sources. Under certain circumstances persistent snowfields or glaciers can be milked during dry years to accelerate their runoff. The results of this research may prove particularly valuable if snowfall can be increased at higher elevations through cloud seeding.

"Project Skywater" is a Bureau of Reclamation long range research program to increase water supplies by increasing precipitation over watershed areas. Begun in 1962 and originally limited to the Western States, the program has since expanded to a nationwide effort.

The goal of Project Skywater is to develop by 1985 a practical technology for increasing precipitation by 10 to 20 percent. Most of this ambitious program is being carried out under contract by research organizations, colleges and universities, State offices, and Federal agencies such as the Geological Survey, the Weather Bureau, the Forest Service and Soil Conservation Service of the Department of Agriculture, and the Naval Weapons Center. Under the program precipitation modification techniques will be tested over large areas of the country for a number of years to realistically verify and evaluate results.

Television has proven a most effective medium for disseminating weather information. Except in emergency situations, however, Weather Bureau personnel do not make telecasts. The Bureau does supply all the necesary information and forecasts and has trained station personnel in various cities to make weather presentations. In 1961 the American Meteorological Society, a national scientific and professional organization, began issuing a "seal of approval" to those television weathermen whose work meets professional standards.

Incorporated in 1920, the American Meteorological Society today has nearly 9,000 members, with local chapters throughout the 50 States and in many foreign countries. Membership is open to anyone interested in meteorology, with student or associate membership available for nonprofessional members. There is also provision for corporate membership.

The Society successfully met the challenges of demobilization following World War II by providing a placement service for hundreds of war-trained military meteorologists who wished to continue working as weathermen. Fortunately, the growing importance of weather in the planning and operations of many idustries had created a demand for specialized meteorological services and for a new type of weatherman, the private or corporate meteorologist. Many firms have since established their own meteorology departments, while others use the services of commercial meteorological consultants.

The American Meteorological Society and the American Institute of Aeronautics and Astronautics cosponsor annual joint meetings, as well as special events such as the National Conference on Aerospace Meteorology of 1968. The AIAA's technical committee on atmospheric environment promotes the atmospheric sciences and their engineering applications. Each year the Institute presents its Robert M. Losey Award to honor "outstanding contributions to the science of meteorology as applied to aeronautics."

Direct international cooperation between scientists, rather than government officials, began on a formal basis in 1919 with the establishment of the International Research Council and the International Union of Geodesy and Geophysics, the latter consisting of national committees concerned with broad, nongovermental scientific exchanges. The American Geophysical Union was established as the U.S. committee.

Founded in April 1919, the American Geophysical Union promotes the scientific study of problems involving the figure and physics (including meteorology) of the earth, initiates and coordinates research requiring national and international cooperation, and provides for discussion and publication of such studies and research programs. Throughout its history, the Union has had close ties with other scientific organizations, particularly the American Meteorological Society. The two bodies have been holding joint annual meetings since 1930.

One World of Weather
Weather is one of the most truly universal elements of man's physical environment. It has no respect for political boundaries, and one man's weather today is another's tomorrow. Because of this, there has perhaps been more international cooperation in meteorology than in any other field.

The first international meteorological congress was held in 1872, and in 1873 the International Meteorological Organization was formed in Vienna. It was composed of directors of national weather services, assisted by technical experts. The United States was represented by General Myer and scientists from the Army's fledgling national weather service. In 1893 the first international meeting for meteorology ever held outside Europe was convened in Chicago—a source of particular pride to Chief Harrington, Cleveland Abbe, and other Weather Bureau scientists.

The Weather Bureau began exchanging weather data with Russia in 1907 during the reign of Czar Nicholas. Adding this information to reports from China, Japan, the Philippines, Alaska, and Hawaii, Bureau meteorologists could prepare daily weather maps for the whole Northern Hemisphere.

The international exchange of weather information was interrupted by World War I, but revived when the conflict ended. Russian meteorological data was no longer easily obtainable, however, since the United States did not recognize the new Soviet Government. The United States resumed diplomatic relations in 1935, but it was not until after World War II that Russian weather data again became readily available to American forecasters. Since the Second World War there has been consistent meteorological cooperation between the United States and the Soviet Union.

In 1951 the International Meteorological Organization became the World Meteorological Organization (WMO) and began formal affiliation with the United Nations. Dr. Francis Reichelderfer, Chief of the United States Weather Bureau, was elected the first President of the new organization.

Today the WMO acts as the intergovernmental coordinator for international weather services and programs. It assists in setting up weather stations and training meteorological personnel, particularly in underdeveloped countries, and promotes the application of meteorology to the problems of agriculture, aviation, and water resource development.

Sixty-seven nations formally participated in the International Geophysical Year (IGY) of 1957-58, a comprehensive, 18-month global study of the earth's geophysical properties. One permanent result of this ambitious program was the establishment of world geophysical data centers in both the East and West to provide international exchange of scientific information. Among others in the United States, world data centers for seismology, gravity, tsunami, geomagnetism, meteorology, nuclear radiation, ionosphere, airglow, cosmic rays, auroras, and solar observations are currently collocated according to discipline at several of the Environmental Science Services Administration's geophysical data centers.

The success of the IGY led to the international Indian Ocean hydrologic and oceanographic expedition of the early 1960's and the International Years of the Quiet Sun (1964-65), when scientists from 71 countries studied solar-terrestrial relations during a period of weak solar storm (sun spot) activity.

Acting in response to President Kennedy's 1961 call for an intensified cooperative effort in international meteorology, the United Nations requested the World Meteorological Organization and the International Council of Scientific Unions (formerly the International Research Council) to develop a world weather program. The U. N.'s resolution led to two major programs—the World Weather Watch, undertaken by the WMO; and the Global Atmospheric Research Program, planned jointly by the WMO and the Council.

It had long been the dream of weather scientists to develop a permanent, internationally integrated global meteorological observation, communications, and processing system. Technological developments since World War II, particularly the meteorological satellite, has made fulfillment of the dream possible.

Implementation of the World Weather Watch began in 1968 and will take approximately a decade. When it is fully operational, three World Meteorological Centers—at Washington and Moscow in the Northern Hemisphere, and at Melbourne in the Southern Hemisphere—will collect, process, and disseminate worldwide weather observations and prepare analyses and forecasts on a global basis. Regional centers will prepare predictions for more limited regions, while national meteorological centers will be responsible for weather services to their own countries.

A permanent joint planning staff and committee representing the nongovernmental International Council of Scientific Unions and the intergovernmental WMO are charged with coordination of the Global Atmospheric Research Program (GARP)—a comprehensive program to obtain better scientific understanding of the atmosphere's physical and dynamic processes.

The Barbados Oceanographic and Meteorological Experiment (BOMEX), a massive U.S. assault on the secrets of air-sea in-

During an address to the United Nations General Assembly on September 25, 1961, President John F. Kennedy proposed international cooperation in solving the world's weather problems.

teraction, was 'the first of a series of research projects planned under GARP. BOMEX focused on a 90,000 square mile area of the Atlantic east of Barbados in the West Indies where the ocean is deep, its currents converge, and the trade winds blow with a long overwater trajectory. Here 10 ships, 24 planes, and an array of buoys, balloons, satellites, electronic equipment, and approximately 1,500 scientists and technicians sampled and surveyed the ocean, the atmosphere, and their interface during May, June, and July of 1969.

The largest cooperative venture ever undertaken by this country's meteorological and oceanographic communities, BOMEX involved the Air Force, Navy, Army, Bureau of Commercial Fisheries, Coast Guard, NASA's Mississippi Test Facility, the Atomic Energy Commission, and the National Science Foundation. Nongovernment participants included the National Center for Atmospheric Research in Boulder, Colorado, a score of universities, and a number of commercial laboratories. The Environmental Science Services Administration directed and coordinated the project, which recorded trillions of bits of information and generated miles of magnetic computer tape. This massive body of data is currently being digested for analysis.

BOMEX is to be followed by more comprehensive, regional projects. A project to gather a complete set of global environmental data for a full year will be undertaken by the end of the decade.

In the World Weather Watch and GARP, many types of meteorological observations from many types of instrument platforms of many nations will be combined to meet the data requirements for global weather forecasting and for the research necessary to improve it. Present technology can provide the observations and communications needed, but at prohibitive cost. Tomorrow's technology promises more precise observations at significantly lower cost.

Satellite equipment to collect and transmit observations made by instrumented, drifting buoys and balloons, as well as ships at sea, has been successfully tested by NASA.

The Navy developed automatic weather observation buoys for coastal waters during World War II. Subsequently, NOMAD, a larger buoy which could transmit its observations up to 1,000 nautical miles, was developed. In 1964 nuclear power was added, providing an observation platform capable of reporting for a year or more.

The original NOMAD was the first deep sea weather buoy to be anchored successfully in approximately 2,000 fathoms of water. From its location in the Gulf of Mexico, the buoy detected and reported on Hurricane Ethel in 1960 and sent hourly weather and sea surface temperature reports throughout the 1961 passage of Hurricane Carla, one of the most violent hurricanes to strike the Texas coast this century.

The National Council on Marine Resources and Engineering Development has assigned the Coast Guard the task of developing, testing, and evaluating networks of automatic environmental data buoys in the coastal North American deep ocean and estuarine waters and the Great Lakes to provide a better scientific understanding of the marine environment. Results can also be applied to the world weather program.

Balloons are also part of the new technology. At the National Center for Atmospheric Research in Boulder, a new type of balloon is being tested. Unlike standard weather balloons, which rise and expand until they burst at high altitudes, the new *horizontal* sounding balloon is essentially rigid and has only enough helium to take it to a predetermined flight altitude. There the balloon rides the globe-circling winds.

The United States and New Zealand are jointly testing this Global Horizontal Sounding Technique (GHOST) and have launched dozens of such balloons over the Southern Hemisphere. Some have floated around the world for almost a year. NASA is currently developing miniaturized electronic instruments for the GHOST balloons and aiming for a test of a satellite-balloon readout system in the early 1970's.

Other countries are also pursuing technological development programs. Both the U.S.S.R. and France are developing meteorological satellites, while Norway, Germany, and the U.S.S.R. are working on buoys. The United Kingdom and U.S.S.R. are developing remote sensing devices for satellites, and many nations are perfecting automated weather stations. In addition, in a cooperative venture, France is working on a balloon and satellite system which it plans to test in the Southern Hemisphere

An artist's conception of the World Weather Watch's environmental observation system when fully implemented.   —ESSA

The World's first atomic powered weather station was installed
on remote Axel Heiberg Island in the Canadian Northwest in
1961. A similar station is in operation near the South Pole.
—ESSA

in 1971. NASA will launch the French-developed Cooperative Applications Satellite in 1970 to track the balloons.

Besides participating in international programs, the Environmental Science Services Administration supports meteorological research and development projects in numerous countries. Many U.S. meteorologists have gone abroad under WMO programs, while scores of foreign weathermen come to the United States for training on WMO fellowships. In addition, the United States provides some technical assistance through the Agency for International Development.

Toward the Third Millennium
The television eyes of weather satellites already provide a space view of earth's weather systems. With astronauts landing on the moon and interplanetary travel being considered, meteorologists are increasingly involved in the Nation's space program.

Big rockets are extremely vulnerable to weather during transfer to the launch pad, countdown, and launch, with lightning and strong winds the elements of major concern. The launch itself can be delayed by low clouds or poor visibilities. And when the spacecraft and its passengers plunge back through the atmosphere to splashdown and recovery, weather is once again a critical factor.

The Weather Bureau's Space Operations Support Division provides the climatological studies and weather forecasts needed by civilian space programs, whether for test projects or manned space flight. At NASA's Mississippi Test Facility, for example, Weather Bureau meteorologists forecast atmospheric sound-propagation conditions before static firings of the big, noisy Saturn engines are made, to keep rattled windows and nerves in nearby towns to a minimum. In the manned spaceflight program, Gemini V was brought down one orbit earlier than planned to avoid its landing near the tropical storm that became Hurricane Betsy.

There are five spaceflight meteorology offices. The Cape Kennedy unit provides forecasts and briefings during prelaunch and launch operations, while the Houston section provides briefings for the orbital and recovery mission phases. The Suitland (Maryland) section, collocated with the National Meteorological Center and National Environmental Satellite Center, produces nearly worldwide weather analyses and forecasts. The Miami group, collocated with the Weather Bureau's National Hurricane Center, contributes analyses and forecasts for the Gulf of Mexico, Southeastern United States, and Atlantic Ocean areas. The Honolulu representative provides specialized North Pacific forecasts, which are coordinated with the Navy when they involve the deployment of the Pacific recovery forces.

ESSA's Space Disturbance Laboratory in Boulder, Colorado, provides another kind of "weather" service. Linked with observing stations around the world, Laboratory scientists constantly monitor the sun's activity and issue daily warnings of solar flares, magnetic storms, and other space disturbances. Solar flares often emit streams of charged particles which could subject astronauts and space experiments to high radiation levels.

The practical application of solar flare forecasting goes beyond protecting our astronauts. Magnetic storms generated by these flares or by other less spectacular solar disturbances can disrupt radio communication for extended periods. The storms also produce voltage surges in long distance power lines and the Laboratory's warnings help power company managers prevent unnecessary interruptions in service.

ESSA, the Air Force, and NASA operate a cooperative global ring of observatories to monitor the face of the sun. These stations are part of a larger international network of cooperative solar observatories. The Air Weather Service's Solar Forecast Center in Colorado provides tailored forecasts for Department of Defense activities. Navy Research Laboratory scientists, concerned with the effects of solar flares on fleet telecommunications, are also studying the sun.

In 1954, Dr. John von Neumann urged the Weather Bureau to undertake theoretical studies of atmospheric circulation. The effort began in October 1955 as the General Circulation Research Section of the Weather Bureau. With the formation of ESSA this became the Geophysical Fluid Dynamics Laboratory, now located at Princeton University.

The Laboratory's computer models are used to simulate global atmospheric circulation, and its research is aimed at achieving the maximum understanding possible of atmospheric processes. Recent computer models concern the sea as well as

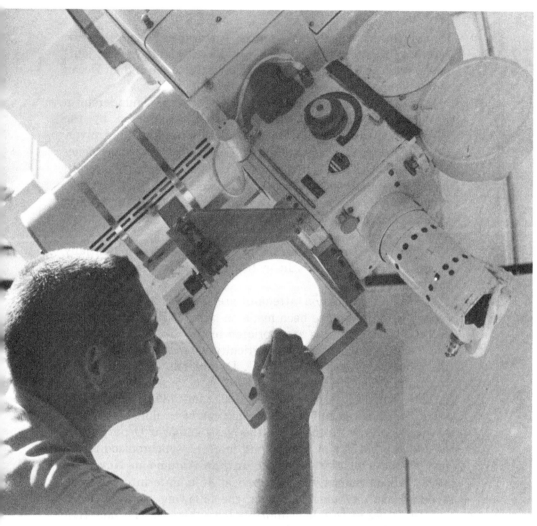

An Air Force weatherman uses a solar telescope to chart
solar-flare patterns. —U.S. Air Force

the air, and a multilayer model of the ocean is being developed
in the hope of eventually synthesizing the mathematical equiva-
lent of both air and sea—the earth's entire fluid mantle.

Following a series of devastating hurricanes, Congress in 1955
made the Weather Bureau responsible for a national hurricane
research project. The Air Force originally provided the aircraft
needed to study the giant storms, but in 1960 the Weather
Bureau acquired three of its own planes. Today, this research
flight group numbers four aircraft and is one of 13 major components of ESSA's Research Laboratories. Its planes chase

thunderstorms when there are no hurricanes to hunt, probe severe winter storms along the Great Lakes, and have even journeyed halfway around the globe to investigate the Indian Ocean monsoon circulation.

Postwar severe storm research began with the Thunderstorm Project in 1946-47, an exhaustive study conducted by the Weather Bureau, the military weather services, and the University of Chicago. Thunderstorms in Florida then Ohio were penetrated, probed, and exhaustively analyzed by radar, aircraft, upper air soundings, and a dense network of ground stations. Meteorologists learned much about the structure, turbulence, mechanisms, and power of thunderstorms from this now-classic study.

Today weather modification is one of the most potentially promising research fields in meteorology. It is also one of the most controversial.

Weather modification attempts are nothing new, though in recent years they have been made on a more scientific basis. As early as 1890 Congress appropriated money to see if explosions could cause rain. Secretary of Agriculture Jeremiah Rusk turned the money over to General R. G. Dyrenforth, who was chosen to conduct the experiments.

The General's tests, made in 1891 and 1892, consisted of sending explosives and slow-burning time fuses aloft on balloons or kites. Later, with the aid of the Navy and War Departments, he added a battery of cannon and mortars, synchronized to fire just as the explosives went off. In San Antonio in November 1892, balloon bombs were released at 45-minute intervals from noon of one day until 3 o'clock the following morning, when indeed, it began to rain. Unfortunately, there was no rain (nor seldom even clouds) associated with subsequent attempts, even though as each successive test failed, more artillery and explosives were added; the sole result was an increasingly irritated and insomniac citizenry.

During the 1930's, several European scientists theorized that rain could be caused by seeding clouds with dry ice and supercooled water ice. In 1946 Irving Langmuir and Vincent Schaefer of General Electric succeeded in modifying cloud structures by seeding them with dry ice, and a little later Bernard Vonnegut of GE showed that silver iodide crystals would produce the same result. By 1951 scientists of 30 nations, including the United States, were actively experimenting in the field of weather modification.

Controversy concerning the results of these experiments led to the creation of a presidential advisory committee in 1953 to evaluate modification attempts and explore the need for regulatory controls. After analyzing the results of 427 tests conducted both in the United States and France, the committee concluded that cloud seeding in the mountains of the western United States had probably increased precipitation by 10 to 15 percent, but that elsewhere there was little evidence of increased rainfall. Even the increase in the mountains was challenged by some meteorologists and statisticians.

On January 18, 1966, the National Academy of Sciences Panel on Weather Modification, after several years of investigation and debate issued a report which stated in part:

> The available evidence, though not conclusive, indicates that artificial nucleation [seeding] techniques, under certain meteorological conditions, may be used to modify the space or time distribution of precipitation. Specifically, we find some evidence for precipitation increases of as much as 10 or even 20 percent over areas as large as 1,000 square miles over periods ranging from weeks to years.

The Panel recommended the undertaking of carefully designed experiments to evaluate the effects of seeding on various types of storms, including comprehensive exploration of hurricane and hail storm modification techniques. Regarding large scale weather or climate changes, however, the Panel declared, "categorically . . . there is, at present, no known way deliberately to induce predictable changes."

Beyond the fundamental question of whether man can significantly alter atmospheric processes, there is a maze of legal, social, and economic problems that also must be resolved—such as liability for weather damage in the vicinity of seeding activities or for apparent unfavorable weather changes downstream—whether in the next field, county, State, or nation.

Because of these problems, many States have enacted weather modification legislation. Some require licensing of commercial operators or registration of each project before it begins, while in Pennsylvania the individual counties have the rights to ban weather modification projects. In 1966 the National Science Foundation issued a regulation requiring 30 days' notice of weather modification attempts by both commercial and Government agencies, in an attempt to coordinate the ever-increasing, sometimes overlapping activity in this field.

ESSA's Research Loboratories conduct a number of weather modification projects, investigating both the possibilities and consequences of modifying rain- and snow-producing cloud masses.

A current cloud and precipitation project began in 1968 as a joint ESSA-U.S. Navy operation. In these early experiments, heavy silver iodide seeding caused selected tropical cumulus clouds over the Caribbean to grow thousands of feet taller than similar, unseeded clouds nearby; the taller clouds are more likely to produce rain or heavier rain. In the spring of 1969 ESSA seeded cumulus clouds over the Florida Everglades, where any effects on local precipitation could be more easily measured. Similar experiments are planned for the future.

The lake-effects snow modification project, a contractual effort with Buffalo's Cornell Aeronautical Laboratory, the State University of New York colleges at Albany and Fredonia, and Pennsylvania State Univeristy, has involved limited cloud seeding experiments around Lakes Erie and Ontario. The purpose of these experiments is to spread further inland some of the heavy, crippling snowfall dumped on a narrow 20-mile strip along the Lakes' shoreline in late autumn or early winter, when arctic air flowing southeastward over the still unfrozen water picks up heat and moisture and produces local snowstorms over the downwind shores. This condition may persist for several days and accumulations can exceed 3 feet per day.

Hurricanes are the largest game available to the cloud seeder. Their power is awesome. Dr. Joanne Simpson of ESSA's Atmospherics and Chemistry Laboratory has calculated that a mature hurricane of moderate size and intensity releases as much condensation heat energy in a single day as the fusion energy of roughly 400 hydrogen super-bombs. To weaken such a colossus, man must exploit the storm's internal instabilities.

A Gulfport, Mississippi victim of Hurricane Camille.　　　—ESSA

Camille also beached this fishing trawler in Biloxi, Mississippi. —ESSA

Project Stormfury, a cooperative undertaking of the Department of Commerce and the Navy assisted by Air Force and university scientists, was established in 1962 and is under the direction of ESSA's National Hurricane Research Laboratory. It involves two modification techniques. In eyewall experiments, Navy aircraft seed the towering clouds surrounding the hurricane eye with silver iodide to encourage liquid water droplets present at below freezing temperatures to freeze. This change of state releases heat energy into the storm, and theoretically should decelerate its worst winds.

Another experiment involves seeding hurricane rain bands. The idea is to generate convective currents which will force low-level incoming air directly upward into the storm's "exhaust system," starving the storm's thermal engine.

Hurricane Esther was seeded in 1961, before Stormfury was organized, and Hurricane Beulah, in 1963. Through 1968 no other storm provided the conditions of position, development, and path necessary to initiate Stormfury. Seeding of Esther and Beulah produced some encouraging results. The seeded portion of Esther's eyewall faded from a monitoring radarscope, indicating either the change of liquid water to ice crystals or the replacement of large droplets by much smaller ones. And soon after Beulah was seeded, the central pressure of the eye rose and the area of maximum winds moved away from the storm center. In August 1969 Hurricane Debbie was seeded; the results are being analyzed as this is written.

In addition to its Stormfury efforts, the National Hurricane Research Laboratory and a number of universities are attempting to develop mathematical models for computer simulation of the life cycle of a hurricane. This would provide a means of modifying a model hurricane and of predicting the effects of such modification before it was actually attempted in the atmosphere.

From a weather-modification viewpoint, thunderstorms and tornadoes are a bigger problem than hurricanes. Severe thunderstorms are too violent to risk routine airborne penetration, while tornadoes are simply incompatible with flight. Another difficulty is the short life span of these storms. It is rare that a meteorologist can be in the right place at the right time, and almost impossible to place instruments to measure such elements as the actual wind speed in a tornado funnel—which is still unknown. Despite these handicaps, researchers of ESSA's Severe Storms Laboratory in Norman, Oklahoma and other meteorologists continue to dog squall lines and thunderstorms with mobile equipment to learn as much as they can about these storms, so that eventually modification theories may be tested in the atmosphere.

Hail is a product of severe thunderstorms that destroys about $300 million worth of crops and property in the United States per year. Any practical hail suppression technique would have to begin before the thunderstorm was well organized to successfully prevent the growth of large hailstones. This introduces the difficult problem of identifying hail-producing storms before they produce significant hail.

Many countries have attempted to at least limit the size of hailstones. The Russians, for example, use radar-directed artillery shells and rockets to seed severe thunderstorms and report millions of rubles saved each year in reduced crop damage. Other attempts, however, have proven less successful, and an operational hail modification program must await the results of further research and extensive field experiments.

Lightning suppression is another attractive possibility. The objective is to prevent the electrical potential—the difference between cloud and ground electrical charge—from building up to the high levels where the destructive spark of a lightning bolt is generated. Experiments of a joint U.S. Army-ESSA project being conducted near Flagstaff, Arizona indicate that a thunderstorm's electrical charge may be short-circuited before it reaches such levels. In the experiments, strands of metalized nylon thread are distributed in the clouds in an attempt to dissipate the electrical field harmlessly through local discharges from each fiber. (In the Department of Agriculture's Project Skyfire, the Forest Service is attempting to increase the number of snow crystals in thunderstorm clouds by seeding, with each crystal acting as a conductor for electrial discharge.)

The death and destruction wrought by lightning makes these efforts worthwhile. In the United States alone, lightning annually causes 150 deaths, 250 injuries, and $100 million damage.

One advance in at least small scale weather modification came out of World War II. It was called FIDO, for "Fog, investigations dispersal of." During the war, Allied planes in the British Isles were frequently grounded by fog. The British found that by burning hundreds of gallons of fuel oil along the runway, the fog could be dissipated. The heat raised the air temperature, causing the suspended water droplets to evaporate.

Fog is generally classified as cold or warm. Cold fogs are composed of water droplets cooled below freezing, and striking success has been achieved in clearing airport approach zones and runways of such supercooled fog and low-hanging clouds by seeding techniques. In pioneering work conducted by the Army, United Airlines, and the Air Force, fog seeding changed the supercooled water droplets to ice crystals, and the fog simply snowed itself out. This type of weather modification is now routine at many of our Northern airports in winter, and at a number of Air Force bases in this country and in Europe.

Warm fog poses a more difficult problem, for it consists of water droplets which are not supercooled. This is the type of fog dispersed by FIDO. Unfortunately, FIDO was a very expensive operation—too expensive for routine peacetime use. Various seeding techniques are currently being tried, but are still in the experimental stage.

During the 1970's a full global environmental observation system will become available to man for the first time in his history. This development, plus the expected fruition of much of today's still embryonic technology and research, will fix the shape and character of the national weather service as we approach a new century.

By the year 2000 a nearly "stationary" satellite some 22,000 miles above the earth's Equator should be monitoring North America's cloud cover with color television by day and infrared sensors by night. This spacecraft, and sister satellites circling the globe at much lower altitude, should also monitor solar radiation and periodically interrogate GHOST balloons in the earth's atmosphere, buoys in her oceans, radar networks, and river and tide gages. Meteorological observations from thousands of weather stations, ships at sea, and islands and aircraft

Composite sequence of an aerial fog seeding operation at Salt
Lake City Airport, using crushed dry ice. The first 3 photos
were taken 5 minutes apart, the lower right one 10 minutes
later.   —UNITED AIR LINES

would be assembled at meteorological centers in most countries of the world and beamed to the satellite sitting over the Equator, which would relay the information almost continuously to the National Meteorological Center and other key weather centers in the United States.

Far below this spacecraft, improved models of smaller ESSA satellites would be orbiting the earth each hour. Besides mapping global cloud cover by both stored and APT cameras and infrared radiometers, they would also carry a future generation of SIRS sensors to sound the vertical variation of temperature, moisture, and other atmospheric elements. In addition, new sensors capable of measuring oceanographic features such as sea state, water temperature, and large-scale current movement may be available.

The information gathered by these satellites would also flow almost continuously to the computers of the National Meteorological Center, helping to produce a continuous, ever-changing mosaic of the earth's weather. From the computers would come weather analyses and forecasts for instant relay to the National Hurricane Center, the National Severe Storms Center, local weather offices throughout the country and, via the geostationary satellite, the other nations of the world.

The weather forecasts of the year 2000 should be considerably longer range than today's, as well as more accurate and more tailored to national needs. They will not be perfect; they never will be.

As global population pressure intensifies and man looks to the sea to supplement his food supply, the weather and climate of the oceans may become as important as the weather and climate of today's farmland. Overhead, with supersonic flight commonplace, we will need aviation forecasts over global distances and to altitudes which today only rockets can reach. In space flight, as interplanetary travel increases, we will have to forecast solar storms more accurately and for longer periods of time than is possible today.

By the time we reach the threshold of the 21st century, meteorologists may know if meaningful weather modification and control is possible; perhaps they will have already developed techniques to free man from some of the environmental hazards and limitations to which he has been subject since his appearance on earth.

Finally, in international meteorology—particularly the World Weather Watch—man may at last find the prototype of global concern and cooperation that he has so long pursued in the political and social spheres.

METEOROLOGICAL MILESTONES: CHRONOLOGY OF THE AMERICAN WEATHER SERVICES

1644–1970

1644–45 The first weather observations in the New World are made by the Reverend John Campanius at the Swedes' Fort, near the present site of Wilmington, Delaware.

1743 Benjamin Franklin deduces the northeastward movement of a hurricane from eclipse observations at Philadelphia and Boston. This is the first recorded instance in which the progressive movement of the storm system as a whole is recognized.

1776–78 Thomas Jefferson at Monticello and James Madison (not the president) at Williamsburg, Virginia, take the first known simultaneous weather observations in America.

1780–92 Four American stations participate in the Mannheim Meteorological Society's 39-station international network of weather observation posts.

1804–06 Lewis and Clark expedition makes regular weather observations.

1814 The Surgeon General issues orders for weather observations at Army posts.

1817 On April 29, Josiah Meigs, Commissioner of the General Land Office, requests local land office registrars to take three daily weather observations in the interest of agriculture.

1825 The Board of Regents of the State of New York establishes a State climatological network at the 30 academies under its control.

1830   James P. Espy announces his convective theory of the formation of clouds, whereby rising air is cooled by expansion, causing condensation of water vapor and thus clouds.

1831–46 William C. Redfield publishes a long series of papers showing that hurricanes are great revolving storms with the winds rotating around the storm's center. He also traces hurricane tracks from the West Indies to the East Coast of the United States.

1836   Elias Loomis publishes his work on storm dynamics and the path of storms, the first of a long series of important meteorological papers.

1838   The Pennsylvania State Legislature makes the first appropriation of public money for weather services in the United States. Grants $4,000 to the Franklin Institute of Philadelphia to establish meteorological stations in each county of the State.

The first synoptic weather map in the United States is prepared by Elias Loomis and subsequently published in 1841.

1841   James P. Espy publishes his *Philosophy of Storms*, stressing the importance of the expansion of rising air in the formation of thunderstorms.

1842   Espy appointed the first official U.S. Government meteorologist.

1845   On April 1 the first line of the Morse telegraph— between Washington, D.C., and Baltimore—is opened to the public.

1846–48 Lieutenant Matthew Fontaine Maury publishes the first of his marine temperature, wind, and current charts.

1849   Joseph Henry inaugurates the Smithsonian Institution's telegraphic network of weather observers "to solve the problems of American storms."

1853   The International Maritime Conference is held in Brussels. Beginning of organized marine meteorology.

1853–55   Ebenezer E. Merriam, "The Sage of Brooklyn Heights," publishes a series of local weather forecasts in the New York daily newspapers based largely on telegraphic weather reports also appearing daily in the press.

1856   William Ferrel publishes a mathematical model of the general circulation of global winds and ocean currents. Revised in 1860 and 1889.

1863   Isaac Newton, the newly appointed Commissioner of Agriculture, begins the monthly publication of a bulletin citing weather and crop conditions throughout the country from information furnished by correspondents.

1869   Cleveland Abbe inaugurates a weather reporting and warning service for Cincinnati. On September 1 he publishes his first *Weather Bulletin*; on the 22d, his first forecast.

Increase A. Lapham proposes a Government Storm Warning Service to protect Great Lakes shipping.

1870   On February 2, Congressman Halbert E. Paine introduces a Joint Congressional Resolution requiring the Secretary of War to establish a Government meteorological service. The Resolution is signed by Ulysses S. Grant on February 9.

The *Western Union Telegraph Company's Weather Report*, prepared by Cleveland Abbe, is issued daily (except Sunday) from February 24 to December 10.

The Army Signal Service's new Division of Telegrams and Reports for the Benefit of Commerce begins operations as the Nation's weather service on November 1. On November 8 the first "cautionary storm signal" is issued for Great Lakes shipping by Increase A. Lapham.

1871   Cleveland Abbe makes first official public weather forecasts ("probabilities") on February 19.

Marine meteorological service begins in June when the Signal Service requests three daily simultaneous weather observations be logged by merchant vessels.

1872–81 *Weekly Weather Chronicle* published. Direct ancestor of today's *Weekly Weather and Crop Bulletin* published jointly by the Departments of Agriculture and Commerce.

1873 Daily forecasts for farmers displayed at rural post offices.

A river stage and flood warning service started by the Signal Service.

The International Meteorological Organization, composed of the directors of national weather services, is founded in Vienna. General Myer represents the United States.

Cleveland Abbe starts the *Monthly Weather Review*, a meteorological journal still being published.

1875–84 The Signal Corps publishes a daily *Bulletin of International Simultaneous Meteorological Observations*. In 1878 an international weather map is added. These are not current, but rather historical series, used to study past weather.

1876 James Henry Coffin's *The Winds of the Globe, or the Laws of the Atmospheric Circulation over the Surface of the Earth* is published posthumously. Based on research done in collaboration with the Smithsonian Institution.

Cleveland Abbe and aeronaut S. A. King study the depth of the seabreeze at Coney Island, New York from a moored balloon.

1879 The first newspaper weather map is published in the *New York Graphic*.

1880 Frost warnings are issued for Louisiana sugar growers. Later special weather services are also established for cotton, rice, tobacco, corn, and wheat growers.

1881 The Signal Corps sponsors the organization of supplementary State weather services to aid agriculture. The volunteer observers in these networks number 2,000 by 1892.

The First Polar Year. As part of an international effort, the Signal Corps establishes weather outposts in Alaska and the Northwest Territories. Greeley's expedition to Lady Franklin Bay ends in tragedy; only 7 of 25 members still alive when rescuers arrive in June 1884.

1884   Signal Corps Lieutenant J. P. Finley studies tornadoes and makes practice severe storm alerts.

Abbott Lawrence Rotch establishes Harvard's Blue Hill Meteorological Observatory.

1885   In cooperation with the British Meteorological Office the Signal Corps begins issuing warnings for Atlantic Ocean storms.

1885–90   Professor Henry Allen Hazen makes balloon flights to take meteorological observations.

1887   The Signal Corps' marine meteorological work is transferred to the Navy Hydrographic Office.

1888–89   Professor Hazen develops the first practical self-recording meteorological instruments.

1891   The U.S. Weather Bureau is established in the Department of Agriculture by the transfer of the civilian meteorological service from the Army Signal Corps on July 1.

1893   The first international meteorological conference held outside Europe meets in Chicago.

1894   Meteorological recording instruments sent aloft in kites at Harvard's Blue Hill Observatory.

1895   Experimental kite observations begun by the Weather Bureau under Professor Marvin, who develops the kite meteorograph in 1896.

On September 30, the first *Washington Daily Weather Map* is published by the Weather Bureau. The series is still being published (weekly) by ESSA's Environmental Data Service.

1898   The Weather Bureau begins regular kite observations. The last flight is made in 1933.

The Spanish American War causes concern for U.S. Naval units, leads to establishment of a hurricane warning service.

1900   Cable exchange of weather warnings and information with Europe.

1901–02 The Weather Bureau experiments with wireless communication between Hatteras and Roanoke Island, North Carolina.

1902   The Marconi Company begins broadcasting Weather Bureau forecasts by wireless telegraphy to Cunard Line steamers.

1904   Responsibility for marine meteorological work is transferred to the Weather Bureau from the Hydrographic Office. In turn, the Bureau's work in wireless telegraphy is transferred to the Army and Navy.

1905   The first wireless weather report received from a ship at sea comes from the S.S. *New York* on December 3.

1907   The daily exchange of weather observations with Russia and eastern Asia inaugurated.

1908   Charles Greeley Abbot of the Smithsonian Institution establishes the solar constant, the rate at which radiation from the sun would be received at the earth's surface if there were no atmosphere.

1909   Beginning of the Weather Bureau's current program of free balloon meteorological observations.

1910   The Weather Bureau begins issuing very general weekly forecasts to aid agricultural planning.

1912   As a result of the TITANIC disaster an International Ice Patrol is established, to be conducted by the U.S. Coast Guard.

1913   First fire weather forecast issued. Fire weather service is officially established April 10, 1916.

         Daily radiotelegraphy broadcasts of Weather Bureau marine bulletins by Navy radio stations at Arlington, Virginia and Key West, Florida.

1914   An aerological section is established in the Weather Bureau to meet the growing needs of aviation.

The first daily radiotelegraphy broadcast of forecasts for the farmer is made by the University of North Dakota.

1917   Norwegian meteorologists begin experimenting with the airmass analysis techniques which will revolutionize the practice of meteorology.

1917–18   Marvin meteorographs are mounted on military aircraft in France. In 1925 the Weather Bureau and the Navy initiate daily airplane soundings at Washington, D.C.

1918   The first Signal Corps weather station is established in France in May.

The Weather Bureau begins issuing weather bulletins and forecasts for domestic military flights and for the new air mail routes of the Post Office.

1919   The Naval Aerological Service is established on a permanent basis.

The American Geophysical Union (the U.S. national committee of the International Union of Geodesy and Geophysics) established in April.

1920   The American Meteorological Society is incorporated on January 20.

1921   On January 3 the University of Wisconsin makes a radiotelephony broadcast of weather forecasts, the first systematic use of the new medium. Daily radiophone weather broadcasts followed.

Vilhelm Bjerknes formulates the polar front theory.

1922–28   The Naval Aerological Service reorganizes, adopts airmass analysis techniques.

1926   The Air Commerce Act makes the Weather Bureau officially responsible for weather services to civil aviation.

1927   Carl Gustaf Rossby and the Weather Bureau (with a Guggenheim Fund grant) establish a West Coast prototype for the Bureau's airways meteorological service.

1931   The Weather Bureau begins regular early morning air-
        plane observations at Chicago, Cleveland, Dallas, and
        Omaha on July 1.

1932–33 The Weather Bureau participates in the Second Inter-
        national Polar Year.

1933   The Bureau's last kite station (Ellendale, North Da-
        kota) is closed.

1934   The Weather Bureau establishes an airmass analy-
        sis section. Airmass analysis techniques are officially
        adopted in 1938.

1935   An improved, 24-hour hurricane warning service is
        established.

        C. G. Abbot of the Smithsonian Institution begins
        making long range weather forecasts based on solar
        cycles, a practice he continues today.

1937   The first official Weather Bureau radio meteorograph
        sounding is made at East Boston Airport on August 17.

        Military weather forecasting functions are transferred
        from the Army Signal Corps to the various operating
        units, including the Air Corps, on July 1.

1939   The Weather Bureau initiates automatic telephone
        weather service (in New York City).

1939–40 All military and Weather Bureau airplane observation
        stations are converted to radio meteorograph (radio-
        sonde) soundings.

1939–41 C. G. Rossby and MIT colleagues develop a mathe-
        matical technique to forecast the movement of plane-
        tary waves.

1940   President Roosevelt orders the Coast Guard to man
        ocean weather stations in the Atlantic.

        The Army Air Force and the Navy establish weather
        centers in Washington, D.C.

        The Weather Bureau is transferred from the Depart-
        ment of Agriculture to the Department of Commerce
        on June 30.

        The first official 5-day forecasts are issued from MIT.

1942	The Joint Meteorological Committee of the Joint Chiefs of Staff is established in January to coordinate wartime civilian and military weather activities.
	The Weather Bureau's Analysis Center begins operation early in the year. The Army and Navy weather centers move to adjacent quarters.
1943	Colonel Joseph P. Duckworth and Lieutenant Ralph O'Hair of the Army Air Forces make the first known airplane penetration into the eye of a hurricane.
1945	First fallout forecast for a nuclear explosion made at Alamagordo, New Mexico.
	The Weather Bureau's Extended Forecast Division issues experimental 30-day outlooks for defense industries.
	Army Air Forces Weather Service assumes weather forecasting responsibility for all Army units.
1946	The Navy gives the Weather Bureau 25 surplus aircraft radar to be modified for ground meteorological use.
	International Civil Aviation Organization Conference in September results in a permanent international network of Atlantic weather stations.
	Irving Langmuir and Vincent Schaefer of General Electric succeed in modifying clouds by seeding them with dry ice. First attempt is made on November 13.
1946–47	The Thunderstorm Project. A cooperative research effort of the Weather Bureau, military weather services, and University of Chicago.
1946–48	Fallout forecasting techniques are perfected by the Air Force during nuclear tests at Bikini Atoll and Eniwetok.
1948	Fawbush and Miller, Air Weather Service meteorologists, issue first tornado warning on March 25.
1949	The Air Weather Service's Global Weather Central is established at Offutt AFB, Nebraska.
1950	The Weather Bureau begins issuing 30-day outlooks to the public.

1951   A severe weather warning center begins operation at Tinker AFB in February.

The World Meteorological Organization is established. Weather Bureau Chief Reichelderfer elected first president.

The New Orleans Tabulation Unit moves to Asheville, North Carolina, where it becomes the National Weather Records Center.

1952   The Weather Bureau organizes a severe local storms forecasting unit, which moves to Kansas City, Missouri, in 1954.

1953   The Department of Agriculture begins Project Skyfire, an experimental weather modification program to suppress lightning and prevent forest fires. The project is still active.

A presidential advisory committee is established to evaluate weather modification efforts and the need for regulatory controls.

1954   The Weather Bureau-Navy-Air Force Joint Numerical Weather Prediction Unit is activated at Suitland, Maryland on July 1.

The AN/CPS-9, a weather radar developed for the Air Weather Service by the Army Signal Corps, is unveiled. This is the first radar specifically designed for meteorological use.

1955   Public Law 159 authorizes Federal aid to State and local agencies to combat air pollution. The Department of Health, Education, and Welfare is assigned primary responsibility.

The General Circulation Research Section of the Weather Bureau established. This is now ESSA's Geophysical Fluid Dynamics Laboratory at Princeton University.

1956   The Weather Bureau initiates national hurricane research project.

1957–58 The International Geophysical Year.

1958   The Weather Bureau's National Meteorological Center begins operations in March.

1959   Vanguard II, a satellite developed by the Army's Research and Development laboratories, is launched from Cape Canaveral on February 17. Carrying two photocell units to measure sunlight reflected from clouds, it demonstrates the feasibility of a weather satellite.

The Weather Bureau's first WSR-57 weather surveillance radar is commissioned at Miami on June 26.

The Bureau begins a high altitude weather service for commercial aviation.

A pilot project to provide tailored weather services for agriculture is initiated in the Mississippi Delta.

On October 26, Senator Alexander Wiley of Wisconsin proposes that the United States put into orbit an "international weatherman" satellite to collect weather data and give it out to the entire world.

The Naval Aerological Service becomes the Naval Weather Service.

1960   On April 1 TIROS I, the first fully equipped meteorological satellite, is successfully launched.

In August, NASA and the Weather Bureau invite scientists of 21 nations to participate in the analysis of weather data to be gathered by TIROS II.

Weather Bureau meteorologists at HEW's National Center for Air Pollution Control begin issuing regular advisories of air pollution potential over the Eastern United States. The service is extended to the remainder of the 48 contiguous States in 1963.

1961   In his State of the Union address President Kennedy invites all nations to join with the United States in developing an international weather prediction program.

The Panel on Operational Meteorological Satellites of the National Coordinating Committee for Aviation Meteorology recommends in April that funds be made available for the development of a national operational meteorological satellite system "at the earliest possible date."

The Air Weather Service issues the first official forecast of clear air turbulence (CAT) on November 1.

November 13-22. NASA and the Weather Bureau host an international meteorological satellite workshop to instruct scientists from 27 nations in the techniques of interpreting weather satellite data.

After several years of joint operations with the Air Weather Service, the Weather Bureau assumes responsibility for severe weather forecasting.

4,000 FAA specialists are trained to give weather briefings to pilots as part of a joint FAA-Weather Bureau program to expand aviation weather services.

1962   The Weather Bureau's Project Mercury support group participates in the first U.S. man-in-orbit spaceflight (John Glenn), made on February 20.

On March 16 the Soviet Union launches a satellite which, among other missions, is to investigate "the distribution and formation of cloud pattern." On March 20 Premier Khrushchev emphasizes the importance of a "world weather observation service" using meteorological satellites and urges that the U.S. and the Soviet Union cooperate in such a venture. The U.S. and U.S.S.R. subsequently agree to exchange weather satellite data.

Project Stormfury, a cooperative hurricane modification program, is initiated under the direction of ESSA's National Hurricane Research Laboratory.

The Bureau of Reclamation begins Project Skywater, a long-range weather modification program to attempt to develop the technology to significantly increase precipitation in the United States.

1962–65   The International Indian Ocean Expedition.

1963   The American Institute of Aeronautics and Astronautics is formed by the merger of the Institute of Aerospace Science and the American Rocket Society.

TIROS VIII launched on December 31 with automatic picture transmission (APT) capability.

1964   The Secretary of Commerce establishes the Office of the Federal Coordinator for Meteorology.

1964–65   The International Years of the Quiet Sun.

1965   The Weather Bureau becomes a component of the newly formed Environmental Science Services Administration (ESSA) on July 13. Today, its sister components are the Coast and Geodetic Survey, National Environmental Satellite Center, Environmental Data Service, and ESSA Research Laboratories.

A national natural disaster warning (NADWARN) system is proposed in October, with full implementation to take several years.

The International Hydrologic Decade begins.

1966   The National Academy of Sciences Panel on Weather Modification issues its final report on January 18; finds ". . . some evidence for precipitation increases . . ."

The national operational satellite system is formally established on March 5, when NASA transfers control of ESSA (Environmental Survey Satellite) 2 to the National Environmental Satellite Center. The system involves having two satellites in orbit at the same time, one with picture storage capability, the other with automatic picture transmission (APT) equipment.

Meteorologists from 25 countries meet in London for the first International Clean Air Congress.

1967   The Naval Weather Service is made a separate command on July 1.

Responsibility for issuing air pollution potential advisories is transferred from HEW's National Center for Air Pollution Control to the Weather Bureau's National Meteorological Center on July 7.

1968   Implementation of the World Weather Watch begins.

1969   BOMEX, the first project of the Global Atmospheric Research Program (GARP), is completed off Barbados during May, June, and July.

SUBJECT INDEX
(Entries in parentheses refer to photographs)

Index   204

Hurricane observations radioed from ships at sea, 44
Hurricane rain bands, 181
Hurricane reconnaissance flights, 114, 143, 175, 197 (113)
Hurricane research, 175, 181, 198, 201
Hurricane seeding, 178
Hurricane theory, 190
Hurricane tracks, 190
Hurricane warning service, 23, 114, 146, 196
Hurricanes:
  1960, Ethel, 170
  1961, Carla, 170
  1961, Esther, 181
  1963, Beulah, 181
  1965, Betsy, (123)
  1969, Camille, (143, 179, 180)
  1969, Debbie, 181

Ice, (104)
Ice field charts, Maury, 19
Ice storms, (142)
Icebergs, (109)
Institutes for Environmental Research, ESSA, 137
Instrument shelters, (53)
Interior, Department of the, 161
International Civil Aviation Organization (ICAO), 103, 197
International Clean Air Congress, London (1966), 133, 201
International Conference on Aerial Navigation (1893), 52
International cooperation, 19, 23, 103, 110, 146, 149, 189, 193 (168)
International Council of Scientific Unions, 167
International Geophysical Year (1957-58), 166, 198
International Hydrologic Decade (1965-74), 201
International Ice Patrol, 110, 194 (111)
International Indian Ocean Expedition, 167, 201
International Maritime Conference, Brussels (1853), 190
International Meteorological Conference (1893), 193
International Meteorological Organization (IMO), 165, 166, 192
International observation networks, 189
International Polar Years, 193, 196
International Union of Geodesy and Geophysics, 165, 195
International weather map (1875-84), 23

International Years of the Quiet Sun, (1964-65), 201
Iowa Weather Service, 27

Jacobs, Woodrow C., 158
Japanese balloon bombs, 115
Japanese Navy, 96
Jefferson, Thomas, 14, 189 (12, 13)
Johnson, Lyndon B., 135
Joint Hurricane Warning Central, 114
Joint Meteorological Committee, 116, 197
Joint Numerical Weather Prediction Unit, 131, 137, 198

Kennedy, John F., 124 (168)
Khabarovsk, U.S.S.R., (95)
King, S. A., 51, 192
Kite observations, 51, 52, 53, 54, 193, 196
Kite reelhouse, (53)
Kite sounding, (52)
Kite stations, 53
Korean War, 93, 100
Khrushchev, Nikita, 200

Labor Day hurricane (1935), 44
Lady Franklin Bay, Northwest Territories, 24, 193 (25)
Landsberg, Helmut E., 158
Langley, Samuel, 31 (32)
Langmuir, Irving, 176, 197
Lapham, Increase A., 7, 20, 21, 191 (8)
Legal aspects of weather modification, 177, 178
Lettau, Heinz, 91
Lewis and Clark Expedition, 189
Lifesaving, 101
Lightning-caused fires, 161 (160)
Lightning-caused deaths, 182
Lightning protection, 114
Lightning suppression, 161, 182
Lightships, 146
Lillie, Marvin A., (143)
Lindbergh, Charles, 49
Lisbon, Portugal, 48, 63
Long range forecasting, 39, 71, 138
Loomis, Elias, 16, 190
Losey, Robert M., 75, 82
Luftwaffe, 75

McAdie, Alexander, 100
McDonald, Willard F., 44
McKinley, William, 42 (43)
McMurdo Sound, Antarctica, 101
Madison, James, 189
Magnetic storms, 174

Index   206

INSTITUTIONAL INDEX
U.S. Government Agencies and Bodies

National Center for Air Pollution
Control, 132, 199, 201
National Coordinating Committee
for Aviation Meteorology,
199
National Parks Service, 159
National Science Foundation, 169,
178
National Weather Records Center,
148, 157, 198
Office of Civilian Defense, 114
Office of Emergency Planning, 144
Office of Indian Affairs, 159
Office of the Federal Coordinator
for Meteorology, 158, 201
Post Office, 21, 47, 49, 195
Presidential Advisory Committee on
Weather Modification, 177,
198
President's Science Advisory Board,
117

Soil Conservation Service, 159
Weather Bureau, 27, 28, 193, 201
Analysis Center, 116, 131
Extended Forecast Section
(Division), 69, 137, 197
General Circulation Research
Section, 174, 198
Meteorological Research Division,
68
National Hurricane Center,
143, 174, 186
National Meteorological Center,
173, 186, 201
New Orleans Tabulation Unit,
198
Office of Climatology, 158
Space Operations Support
Division, 173
Works Progress Administration
(WPA), 42, 117

Non-U.S. Government Organizations and Institutions

American Geophysical Union, 165,
195
American Institute of Aeronautics
and Astronautics, 165, 200
American Meteorological Society,
164, 165, 195 (102)
American Rocket Society, 200
British Columbia, Dept. of Lands,
Water Rights Branch, 159
Franklin Institute of Philadelphia,
190
International Civil Aviation
Organization (ICAO), 103,
197
International Council of Scientific
Unions, 167

International Meteorological
Organization (IMO), 192
International Union of Geodesy and
Geophysics (IUGG), 165, 195
National Academy of Sciences,
Panel on Weather
Modification, 201
National Center for Atmospheric
Research (NCAR), Boulder,
Colo., 169, 170
Smithsonian Institution, 16, 24, 190,
192, 194
World Meteorological Organization
(WMO), 165, 167, 198

For Product Safety Concerns and Information please contact our EU representative GPSR@taylorandfrancis.com Taylor & Francis Verlag GmbH, Kaufingerstraße 24, 80331 München, Germany

Printed and bound by CPI Group (UK) Ltd, Croydon, CR0 4YY

01/05/2025

01858566-0001